Springer Geophysics

The Springer Geophysics series seeks to publish a broad portfolio of scientific books, aiming at researchers, students, and everyone interested in geophysics. The series includes peer-reviewed monographs, edited volumes, textbooks, and conference proceedings. It covers the entire research area including, but not limited to, geodesy, planetology, geodynamics, geomagnetism, paleomagnetism, seismology, and tectonophysics.

More information about this series at http://www.springer.com/series/10173

Jaroslav Klokočník · Jan Kostelecký
Aleš Bezděk

Gravitational Atlas of Antarctica

Springer

Jaroslav Klokočník
Astronomical Institute
Czech Academy of Sciences
Ondřejov
Czech Republic

Aleš Bezděk
Astronomical Institute
Czech Academy of Sciences
Ondřejov
Czech Republic

Jan Kostelecký
Research Institute of Geodesy, Topography and Cartography
Zdiby
Czech Republic

and

Faculty of Mining and Geology
VŠB—Technical University of Ostrava
Ostrava
Czech Republic

Springer Geophysics
ISBN 978-3-319-56638-2 ISBN 978-3-319-56639-9 (eBook)
DOI 10.1007/978-3-319-56639-9

Library of Congress Control Number: 2017939602

Printed on acid-free paper

This Springer imprint is published by Springer Nature
The registered company is Springer International Publishing AG
The registered company address is: Gewerbestrasse 11, 6330 Cham, Switzerland

If at first the idea is not absurd,
then there is no hope for it

Albert Einstein

Acknowledgements

This work has been prepared in the frame of projects GACR #13-36843S (Grant Agency of the Czech Republic) and RVO #67985815 (Astronomical Institute, Czech Academy of Sciences), partly supported by the project LO 1506 (PUNTIS) from Ministry of Education of the Czech Republic. We thank B. Bucha for computations of the gravitational signal, H.D. Pritchard for his help with the bedrock topography data, and V. Cílek for fruitful discussions about Antarctica. We also thank Ms Johanna Schwarz, senior publishing editor, and other "Springer people" for their care about our manuscript.

Contents

Abstract

The gravity anomalies/disturbances, Marussi tensor of the second derivatives of the disturbing geopotential, represented by selected gravity field models (like EIGEN 6C4 or RET 14), the gravity invariants and their specific ratios, the strike angles, and the virtual deformations (shortly the "aspects") are computed and plotted for Antarctica together with the bedrock (under the ice) topography of Antarctica (BEDMAP 2). First, global views of the whole continent for all the aspects are presented, and then many more details in latitudinal-longitudinal segments and zooms for selected regions and locations are shown as examples of the aspects. The Atlas provides a new, unusual view on Antarctica and may serve the users (geo-scientists and others) to help in their various interpretations.

Keywords Gravitational field of the Earth · Gravity/gravitational field models · Gravity anomalies/disturbances · Marussi tensor · Gravity invariants · Strike angles · Virtual deformations · Antarctica · Bedrock topography of Antarctica

Chapter 1
Introduction

Gravity maps of Antarctica were recently issued as a part of the World Gravity Map (Bonvalot et al. 2012); they contain maps with the free-air, Bouguer and isostatic anomalies. They are based on the EGM 2008 global combined gravity field model (Pavlis et al. 2008a, b, 2012). In the meantime, new gravity field models using new satellite data (mainly from the GOCE mission, Sect. 3.1) have been created, e.g. EIGEN 6C4 (Foerste et al. 2014) and also the new bedrock topography of Antarctica called BEDMAP 2 (Fretwell et al. 2013), a combination of EIGEN 6C4 and this BEDMAP 2, valid for Antarctica, known as RET 14 (Hirt et al. 2016), has been published. This inspired us to make a better and more comprehensive "gravitational map" of Antarctica. Moreover, we work not only with the gravity anomalies (or disturbances), but also with the second derivatives of the disturbing gravitational potential (the so-called Marussi tensor), with the invariants of the gravitational field, their specific ratio, the strike angle and the virtual deformations. These quantities used together provide much more information than mere gravity anomalies (Chap. 2). [Trying to find one name for all of these functions of the disturbing geopotential, we suggested "derivates", but perhaps the word "aspects" sounds better.]

The Atlas in your hands provides a new information never published elsewhere in this form, only the authors of this Atlas published scientific papers about the use and applications of the various gravitational aspects with EGM 2008, later with EIGEN 6C4 and now with RET 14 for selected parts of the world and specific purposes (Klokočník et al. 2010, 2013, 2016; Klokočník and Kostelecký 2014).

The user of this Atlas may find it complicated to obtain a detailed knowledge about the input data used to compute the aspects and to plot the presented figures (this comment applies to BEDMAP 2 but also to other data sets). These models still suffer from a certain degree of non-homogeneity due to the irregular data distribution and varying quality and they show various spatial resolution, precision and reliability in different locations. It is recommended to contact the authors of these models before interpreting results.

J. Klokočník et al., *Gravitational Atlas of Antarctica*, Springer Geophysics,
DOI 10.1007/978-3-319-56639-9_1

It is not difficult to interpret the gravity anomalies or the second radial derivative of the potential geometrically; it is more difficult to imagine the strike angles or virtual deformations, and it is very difficult to understand what the gravity invariants mean. This is relatively or completely new in applications and we all are learning.

For various geoapplications we always need a close cooperation with geoscientists (geologists, geophysicists, morphologists, oceanographers, glaciologists and others). Gravity data are very useful but yield often only a boundary conditions to solve diverse problems in geoscience. The Atlas is not intended to interpret anything directly but as an inspiration for people who will use it for their own interpretations. The reader is invited to contact the authors of this Atlas to compute and plot the aspects for the areas of his/her interest.

References

Bonvalot S., Balmino G., Brias A., Kuhn M., Peyrefitte A. and Vales N. 2012. World Gravity Map. Bureau Gravimetr. Interntl., map, edited by CGMW-BGI-CNES-IRD, Paris.

Förste Ch., Bruinsma S., Abrykosov O., Lemoine J-M. et al. 2014. The latest combined global gravity field model including GOCE data up to degree and order 2190 of GFZ Potsdam and GRGS Toulouse (EIGEN 6C4). *5th GOCE user workshop*, Paris 25–28, Nov.

Fretwell P, Pritchard H, Vaughan D, Bamber J, et al. 2013. Bedmap2: Improved ice bed, surface and thickness datasets for Antarctica. *The Cryosphere*, 7, 375–393, doi:10.5194/tc-7-375-2013

Hirt Ch., Rexer M., Scheinert M., Pail R., Claessens S. and S. Holmes 2016. A new degree-2190 (10 km resolution) gravity field model for Antarctica developed from GRACE, GOCE and Bedmap 2 data, *J. Geod.* 90: 105–127; doi:10.1007/s00190-015-0857-6.

Klokočník J., Kalvoda J., Kostelecký J., Eppelbaum L.V. and Bezděk A. 2013. Gravity Disturbances, Marussi Tensor, Invariants and Other Functions of the Geopotential Represented by EGM 2008, *ESA Living Planet Symp.* 9–13 Sept. 2013, Edinburgh, Scotland. Publ. in: August 2014: *J. Earth Sci. Res.* 2: 88–101.

Klokočník J. and Kostelecký J. 2014. Gravity signal at Ghawar, Saudi Arabia, from the global gravitational field model EGM 2008 and similarities around, *Arab. J. Geosciences*, on-line, doi:10.1007/s12517-014-1491-y; ISSN 1866-7511, Springer-Verlag.

Klokočník J., Kostelecký J., Bezděk A. and Pešek I. 2016. On feasibility to detect volcanoes hidden under the ice of |Antarctica via their "gravitational signal", *Annals of Geophysics* 59 (5), S0539. http://dx.doi.org/10.4401/ag-7102

Klokočník J., Kostelecký J., Pešek I., Novák P., Wagner C. A. and Sebera J. 2010. Candidates for multiple impact craters? Popigai and Chicxulub as seen by the global high resolution gravitational field model EGM08. *EGU Solid Earth* 1: 71–83, doi: 10.5194/se-1-71-2010

Pavlis N. K., Holmes S. A., Kenyon S. C. and Factor, J. K. 2008a. An earth gravitational model to degree 2160: EGM 2008. *EGU General Assembly* 2008, Vienna, 13–18 April.

Pavlis N. K., Holmes S. A., Kenyon S. C. and Factor, J. K. 2008b. EGM 2008: An overview of its development and evaluation. National Geospatial-Intelligence Agency, USA, conf.: *Gravity, Geoid and Earth Observation 2008*, Chania, Crete, Greece, 23–27 June.

Pavlis N. K., Holmes S. A., Kenyon S. C. and Factor J. K. 2012. The development and evaluation of the Earth Gravitational Model 2008 (EGM 2008). *J. Geophys. Res.* 17: B04406, doi:10.1029/2011JB008916, 2012.

Chapter 2
Theory

2.1 Definitions

The theory presented here comes mainly from Pedersen and Rasmussen (1990), Beiki and Pedersen (2010) and from our recent papers Kalvoda et al. (2013) and Klokočník et al. (2014, 2016). A short outline of the theory for a convenience of the reader (non-specialist in physical geodesy) follows.

The disturbing static gravitational potential outside the Earth masses in the spherical harmonic expansion is given by the well-known formula:

$$T(r,\phi,\lambda) = \frac{GM}{r}\sum_{l=2}^{\infty}\sum_{m=0}^{l}\left(\frac{R}{r}\right)^{l}\left(C'_{l,m}\cos m\lambda + S_{l,m}\sin m\lambda\right)P_{l,m}(\sin\phi) \tag{2.1}$$

where GM is a product of the universal gravitational constant and the mass of the Earth (also known from satellite analyses as the geocentric gravitational constant), r is the radial distance of an external point where T is computed, R is the radius of the Earth (which can be approximated by the semi-major axis of a reference ellipsoid), $P_{l,m}$ *(sin* ϕ*)* are the Legendre associated functions, l and m are the degree and order of the harmonic expansion, (ϕ, λ) are the geocentric latitude and longitude, $C'_{l,m}$ and $S_{l,m}$ are the **harmonic geopotential coefficients** (**Stokes parameters**), fully normalized, $C'_{l,m} = C_{l,m} - C^{el}_{l,m}$, where $C^{el}_{l,m}$ belongs to the reference ellipsoid.

The spherical approximation of the **gravity anomaly** Δg (on free air) is computed

$$\Delta g = -\frac{\partial T}{\partial r} - 2\frac{T}{r} \tag{2.2}$$

or one can use the **gravity disturbance** (it is the same as Eq. 2.2 but without the second term, which is usually numerically small). The gravity anomalies/

J. Klokočník et al., *Gravitational Atlas of Antarctica*, Springer Geophysics,
DOI 10.1007/978-3-319-56639-9_2

disturbances are computed from measurements by gravimeters or derived from measurements performed by means of satellite altimetry.

Gravity gradient tensor $\mathbf{\Gamma}$ (the **Marussi tensor**) is a tensor of the second derivatives of the disturbing potential T of the particular gravitational field model. It is assumed that the model is known to a maximum degree L_{max} (see below some details about EGM 2008 and EIGEN 6C4). The second derivatives are expressed in the local north-oriented reference frame (x, y, z), where z has the geocentric radial direction, x points to the north, and y is directed to the west (Pedersen and Rasmussen 1990):

$$\mathbf{\Gamma} = \begin{bmatrix} T_{xx} & T_{xy} & T_{xz} \\ T_{yx} & T_{yy} & T_{yz} \\ T_{zx} & T_{zy} & T_{zz} \end{bmatrix} = \begin{bmatrix} \frac{\partial^2 V}{\partial x^2} & \frac{\partial^2 V}{\partial x \partial y} & \frac{\partial^2 V}{\partial x \partial z} \\ \frac{\partial^2 V}{\partial y \partial x} & \frac{\partial^2 V}{\partial y^2} & \frac{\partial^2 V}{\partial y \partial z} \\ \frac{\partial^2 V}{\partial z \partial x} & \frac{\partial^2 V}{\partial z \partial y} & \frac{\partial^2 V}{\partial z^2} \end{bmatrix} \tag{2.3}$$

Outside of the source masses (in other words, in source-free region), $\mathbf{\Gamma}$ satisfies Laplace's differential equation; it means that the trace of the Marussi tensor is zero. Tensor $\mathbf{\Gamma}$ is symmetric and thus contains **just five** linearly independent components. These can conveniently be computed by means of the formulae in Hotine (1969).

In the local investigations, the tensor components are used to identify and to map the geological contact information, be it the edges of the source targets or the structural/stratigraphic contact information. The horizontal components help to identify the shape and the geological setting of a target body. The quantity T_{zz} is best suited for a target body detection; T_{zz} helps to define isopath/density relationships of a body mass with relation to its geological setting, see, e.g., Saad (2006).

Under any coordinate transformation, the Marussi tensor $\mathbf{\Gamma}$ preserves **three invariants**; here they are labelled I_0, I_1, and I_2:

$$I_0 = trace(\mathbf{\Gamma}) = T_{\mathrm{xx}} + T_{\mathrm{yy}} + T_{\mathrm{zz}}$$

and it is zero outside the masses of the studied body (known also as Laplace's equation). Then

$$I_1 = \left(T_{xx}T_{yy} + T_{yy}T_{zz} + T_{xx}T_{zz}\right) - \left(T_{xy}^2 + T_{yz}^2 + T_{xz}^2\right) = \sum\nolimits_{\{i,j\}\in\{x,y,z\}} \left(T_{ii}T_{jj} - T_{ij}^2\right) \tag{2.4}$$

$$I_2 = \det(\Gamma) = T_{xx}(T_{yy}T_{zz} - T_{yz}^2) + T_{xy}(T_{yz}T_{xz} - T_{xy}T_{zz}) + T_{xz}(T_{xy}T_{yz} - T_{xz}T_{yy}). \tag{2.5}$$

An invariant can be looked upon as a non-linear filter enhancing sources with big volumes (Pedersen and Rasmussen 1990). They discriminate major density anomalies into separate units (the reader will see many examples in Chap. 5).

Pedersen and Rasmussen (1990) showed that the ratio I of $\boldsymbol{I}_1$ and $\boldsymbol{I}_2$ defined as

$$0 \leq I = -\frac{(\boldsymbol{I}_2/2)^2}{(\boldsymbol{I}_1/3)^3} \leq 1 \tag{2.6}$$

lies always between zero and unity for any potential field. If the causative body is strictly 2D, then $I = 0$. Thus the ratio I can be an indicator of two-dimensionality. If $I = 0$ then we have the necessary but not a sufficient condition for two-dimensionality. When the causative body as seen from the observation point looks more and more 3D-like (say a volcano), then I increases and eventually approaches unity (Pedersen and Rasmussen 1990, p. 138).

The second order derivatives and the invariants provide evidence about the details of near–surface (not deep) structures. The Marussi tensor was already used locally (areas of a few kilometres) for petroleum, metal, diamond, groundwater etc. explorations [e.g., Saad (2006), Mataragio and Kieley (2009), Murphy and Dickinson (2009) and many others]. The full Marussi tensor is a rich source of information about the density anomalies providing useful details about the objects located closely to the Earth's surface. This extra information can be used by tensor imaging techniques to enhance target anomalies, as tested for local features (economic minerals, oil and gas deposits, fault location, etc.), see, e.g., Murphy and Dickinson (2009), Saad (2006).

The third derivatives has been studied by Šprlák et al. (2015), but the question is their practical use (noisy values) and also their physical meaning should be clarified. We tested T_{zzz}, but we do not work with it here.

The gradient tensor $\boldsymbol{\Gamma}$ contains information about strikes. Pedersen and Rasmussen (1990) defined the **strike angle** θ (also known as strike lineaments or strike direction) as follows:

$$\tan 2\theta_s = 2\frac{T_{xy}\left(T_{xx}+T_{yy}\right)+T_{xz}T_{yz}}{T_{xx}^2-T_{yy}^2+T_{xz}^2-T_{yz}^2} = 2\frac{-T_{xy}T_{zz}+T_{xz}T_{yz}}{T_{xz}^2-T_{yz}^2+T_{zz}(T_{xx}-T_{yy})} \tag{2.7}$$

within a multiple of $\pi/2$. Provided that the ratio I (2.6) is small, the strike angle may indicate a dominant 2D structure. The strike angle indicates how the gradiometer measurements rotate within the main directions of the underground structures. In other words, if we were able to rotate with the structure in such a way that the elements of the first row and first column of $\boldsymbol{\Gamma}$ would be identically equal to zero, then we would reach the "correct" direction described by θ. For more details see Beiki and Pedersen (2010) or Murphy and Dickinson (2009).

To define the new quantity **"virtual deformation"** (Kalvoda et al. 2013; Klokočník and Kostelecký 2014) (*vd*), an analogy with the tidal deformation was utilized; one can imagine directions of such a deformation due to "erosion" brought about solely by "gravity origin". If there would be a tidal potential T, then the horizontal shifts (deformations) would exist due to it and they could be expressed in the north-south direction (latitude direction) as

$$u_{\Phi} = l_S \frac{1}{g} \frac{\partial T}{\partial \phi} \tag{2.8}$$

and in the east-west direction (longitudinal direction) as

$$u_{\Lambda} = l_S \frac{1}{g \cos \phi} \frac{\partial T}{\partial \lambda} \tag{2.9}$$

where g is the gravity acceleration 9.81 m s^{-2}, l_S is the elastic coefficient (Shida number) expressing the elastic properties of the Earth as a planet ($l_S = 0.08$), ϕ and λ are the geocentric coordinates (latitude and longitude) of a point P where we measure T; the potential T is expressed in [m^2 s^{-2}]. In our case, T is represented by Eqs. (2.1), (2.8) and (2.9). This mechanism is applied to a standard Earth gravitational model, but the real values of the Shida parameters l_S for the Earth's surface (for our purpose) are not known.

The apparatus of mechanics of continuum is applied to derive the main directions of the tension (Brdička et al. 2000). The tensor of (small) deformation $\boldsymbol{E}$ is defined as a gradient of the shift. It holds that

$$E = \begin{pmatrix} \in_{11} & \in_{12} \\ \in_{21} & \in_{22} \end{pmatrix} = \begin{pmatrix} \frac{\partial u_x}{\partial x} & \frac{\partial u_x}{\partial y} \\ \frac{\partial u_y}{\partial x} & \frac{\partial u_y}{\partial y} \end{pmatrix}. \tag{2.10}$$

where x and y coordinates are a) equivalent to the spherical coordinates in a local coordinate system in the respective point or b) are planar coordinates in stereographic projection. The tensor $\boldsymbol{E}$ can be separated into two parts:

$$\boldsymbol{E} = \boldsymbol{e} + \boldsymbol{\Omega} = (\mathrm{e_{ij}}) + (\Omega_{\mathrm{ij}}) \tag{2.11}$$

where $\mathbf{e}$ is the symmetrical tensor and $\boldsymbol{\Omega}$ the anti-symmetrical tensor of deformation, respectively. The symmetrical tensor is:

$$\mathbf{e} = \begin{pmatrix} \mathrm{e}_{11} & \mathrm{e}_{12} \\ \mathrm{e}_{21} & \mathrm{e}_{22} \end{pmatrix} = \begin{pmatrix} \in_{11} & (\in_{12} + \in_{21})/2 \\ (\in_{12} + \in_{21})/2 & \in_{22} \end{pmatrix} \tag{2.12}$$

and the deformation parameters are:

$\Delta = \mathrm{e}_{11} + \mathrm{e}_{22}$ total dilatation
$\gamma_1 = \mathrm{e}_{11} - \mathrm{e}_{22}$ pure cut
$\gamma_2 = 2\mathrm{e}_{12}$ technical cut
$\gamma = (\gamma_1^2 + \gamma_2^2)^{1/2}$ total cut
$a = ½ (\Delta + \gamma)$ major semi-axis of ellipse of deformation
$b = ½ (\Delta - \gamma)$ minor semi-axis of ellipse of deformation
$\alpha = ½\, \mathrm{a\,tan}\, (\gamma_2 / \gamma_1)$ direction of main axis of deformation.

Note that different branches use different terminology for the same or similar quantities.

To illustrate the virtual deformations *vd*, the semi-axes of deformation ellipse a and b are computed in their relative size. The values of l_S are not known, and, therefore, only the main directions of the *vd* (and not its amplitudes) can be shown.

The computations of all quantities (2.2)–(2.12) defined above, sometimes called "gravitational aspects", shortly *aspects*, have been organized by software based on Holmes et al. (2006), or developed by Sebera et al. (2013), Bucha and Janák (2013), and by the authors of this atlas themselves. Various gravity field models and not only for the Earth were applied; here we present results only for Antarctica with the EIGEN 6C4 and RET 14 (see Sect. 3 about the data below). Our computations with EIGEN 6C4 cover the whole planet; the grid used over the whole world is 5 arcmin in latitude and longitude.

2.2 Comments on the Theory

To show how the various aspects (derivates, functions) of the disturbing geopotential, defined in the previous section, behave (decrease) with increasing distance from the source of the causative body (a density anomaly), it is assumed that the source body is a point mass, with a selected value of GM, located on the Earth surface; all masses of the Earth are concentrated in its centre. It is obvious from the definitions in Sect. 2.1 that in this simplified case, some aspects do not exist or are equal to zero. The remaining aspects are shown in Fig. 2.1; here the simplification leads to formulae for the invariants I_1 and I_2, namely: $I_1 = -3\,(GM)^2\,r^{-6}$ and $I_2 = -2\,(GM)^3\,r^{-9}$. A different speed of decrease of the "gravitational signal" with increasing depth of the source mass body for various aspects can be seen. The slowest decrease is for the gravity anomaly (with r^2). The invariants decrease quickly (with r^6 or r^9). It tells us that aspects I_1 or I_2 best describe the density anomaly dominantly at the surface or in shallow depths under the surface while Δg and T_{zz} are related dominantly to deeper structures. Although all these aspects have the same "common

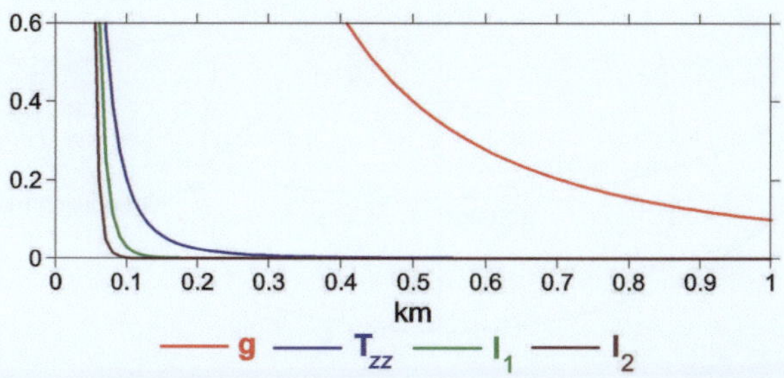

Fig. 2.1 An example of decrease in the values of the aspects of the disturbing gravitational potential with increasing distance (depth) from their source. On the x axis there are kilometres, on the y axis there is an arbitrary quantity (for a simple intercomparison). This is illustrative case for a point mass, representing a density anomaly, with a random value of GM

mother" (a recent gravity field model, Sect. 3.1), they have different properties (behaviour, sensitivity). Thus, the set of the gravitational aspects from Sect. 2.1 provides much more information about the density anomalies than only the traditional Δg, or Δg together with T_{zz}, can do.

A systematic screening of the whole planet Earth was performed. The results of the screening were presented by means of selected examples from regions with varying surface types, high mountain ranges, collision zones of oceanic and continental lithospheric plates, regional fault zones, volcanic chains and large impact craters in Kalvoda et al. (2013), Klokočník et al. (2010, 2014), or Klokočník and Kostelecky (2014), www.asu.cas.cz/~jklokocn.

Geocentric and geodetic latitude. We need to know both types of the "latitude", but all our maps contain geodetic latitude only. The difference between these two quantities is clear from Fig. 2.2; the geocentric latitude ϕ relates to a sphere, the geodetic latitude φ to a reference (rotational) ellipsoid. A relationship between the two latitudes is given by

$$\tan\phi = \left(1 - e^2\right)\tan\varphi,$$

where e is eccentricity of the used reference ellipsoid.

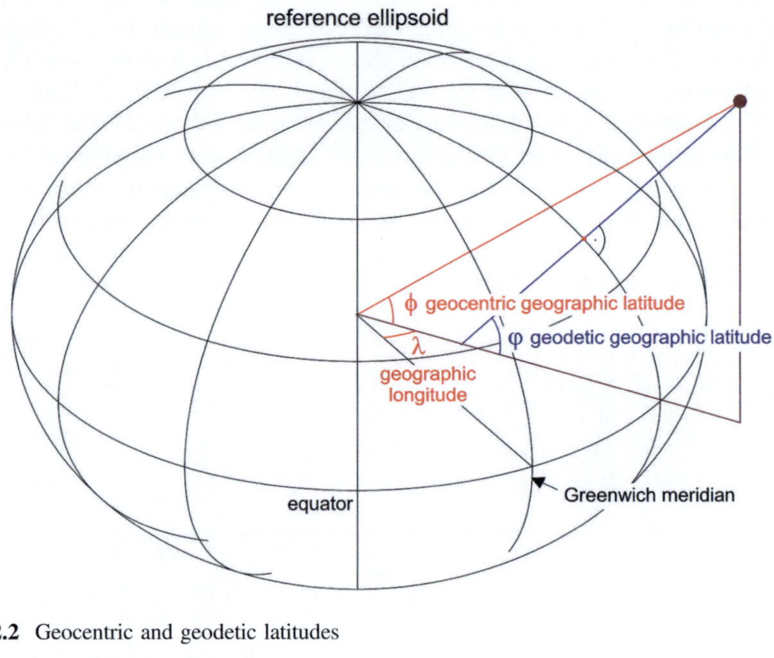

Fig. 2.2 Geocentric and geodetic latitudes

References

Beiki M. and Pedersen L. B. 2010. Eigenvector analysis of gravity gradient tensor to locate geologic bodies. *Geophysics* 75, 137–149, doi: 10.1190/1.3484098.

Brdička M., Samek L. and Sopko B. 2000. Mechanika kontinua (Continuum Mechanics). Academia publ. house, Praha (in Czech language).

Bucha B. and Janák J. 2013. A MATLAB-based graphical user interface program for computing functionals of the geopotential up to ultra-high degrees and orders, *Computers and Geosciences* 56, 186–196. http://dx.doi.org/10.1016/j.cageo.2013.03.012.

Hotine M. 1969. *Mathematical Geodesy*, ESSA, US. Dept. Comm., Environ. Sci. Serv. Admin., Monograph 2, Washington D.C.

Holmes S. A., Pavlis N. K. and Novák P. 2006. *A fortran program for very-high degree harmonic synthesis, version 05/01/2006*, University of New Brunsvick, 2006.

Kalvoda J., Klokočník J., Kostelecký J. and Bezděk A. 2013. Mass distribution of earth landforms determined by aspects of the geopotential as computed from the global gravity field model EGM 2008. *Acta Univ. Carolinae*, Geographica, XLVIII, 2, Prague.

Klokočník J., Kostelecký J., Bezděk A. and Pešek I. 2016. On feasibility to detect volcanoes hidden under the ice of Antarctica via their "gravitational signal", *Annals of Geophysics* 59(5), S0539. http://dx.doi.org/10.4401/ag-7102.

Klokočník J. and Kostelecký J. 2014. Gravity signal at Ghawar, Saudi Arabia, from the global gravitational field model EGM 2008 and similarities around, *Arab. J. Geosciences*, on-line, doi: 10.1007/s12517-014-1491-y; ISSN 1866-7511, Springer-Verlag.

Klokočník J., Kostelecký J., Kalvoda J., Eppelbaum L.V. and Bezděk A. 2014. Gravity Disturbances, Marussi Tensor, Invariants and Other Functions of the Geopotential Represented by EGM 2008. *Journal of Earth Science Research*: 88–101.

Klokočník J., Kostelecký J., Pešek I., Novák P., Wagner C. A. and Sebera J. 2010. Candidates for multiple impact craters? Popigai and chicxulub as seen by the global high resolution gravitational field model EGM08. *EGU Solid Earth* 1: 71–83, doi: 10.5194/se-1-71-2010.

Mataragio J. and Kieley J. 2009. Application of full tensor gradient invariants in detection of intrusion-hosted sulphide mineralization: Implications for deposition mechanisms. *Mining Geoscience, EAGE First Break* 27: 95–98.

Murphy C. A. and Dickinson J. L. 2009. Exploring exploration play models with FTG gravity data. *11th SAGA Biennal Techn. Meeting and Exhib.*, Swaziland, pp. 89–91, 16–18 Sept.

Pedersen B. D. and Rasmussen T. M. 1990. The gradient tensor of potential field anomalies: Some implications on data collection and data processing of maps. *Geophysics* 55: 1558–1566.

Saad A. H. 2006. Understanding gravity gradients–a tutorial, the meter reader. Ed. B. Van Nieuwenhuise, August issue, *The Leading Edge*, 941–949.

Šprlák M. and Novák P. 2015. Integral formulas for computing a third-order gravitational tensor from volumetric mass density, disturbing gravitational potential, gravity anomaly and gravity disturbance, *J. Geod.* 89: 141–157, doi: 10.1007/s00190-014-0767–z.

Sebera J., Wagner C.A., Bezděk A. and Klokočník J. 2013. Short guide to direct gravitational field modelling with Hotine's equations. *J. Geod.* 87: 223–238; http://dx.doi.org/10.1007/s00190-012-0591-2.

Chapter 3
Data

3.1 Recent High-Resolution Gravitational Field Model EIGEN 6C4

The input data to our analysis are the harmonic geopotential coefficients (also known as Stokes parameters) in the spherical harmonic expansion of the disturbing gravitational potential T, namely the numbers C_{lm}, S_{lm} in Eq. 2.1. A set of these coefficients defines a gravitational field model. Today, global gravitational field models of the Earth, based on a variety of satellite and terrestrial data, have a high resolution. *The Earth Gravitational Model* 2008 (EGM 2008, Pavlis et al. 2008a, b, 2012) or *E*uropean *I*mproved *G*ravity model of the *E*arth by *N*ew techniques (EIGEN 6C4, Foerste et al. 2014) are both expanded to degree and order 2190 in spherical harmonics. The number of the terms in such an expansion is roughly $(l + 1)^2$, in this case around 5 million of terms. This expansion corresponds to a spatial resolution of 5 × 5 arc minutes, which is ~9 km half-wavelength on the Earth's surface. For more details on the precision, see Pavlis et al. (2012). Unfortunately, Antarctica is an exception from that limit of 9 km; here the models have a lower resolution, because no terrestrial gravity data are included (although some are available in several areas of Antarctica). It means that for Antarctica, only the satellite gravity data are available (not speaking about topographic data). They are gathered dominantly from the satellite missions GRACE (*G*ravity *R*ecovery *A*nd *C*limate *E*xperiment; e.g., Tapley et al. 2004) and GOCE (*G*ravity field and steady-state *O*cean *C*irculation *E*xplorer; e.g., Floberghagen et al. 2011). Let us mention that ESA's gravity mission GOCE carried the first space gradiometer on board, which measured directly the components of the Marussi tensor $\mathbf{\Gamma}$ (Eq. 2.3, Sect. 2.1). With EIGEN 6C4, the "gravitational signal" is more precise for the majority of parts in Antarctica compared to the signal computed from older models like EGM 2008, which is due to the GOCE data in EIGEN 6C4.

Polar gap. The satellite data in the gravitational models are not available over the whole globe but have small polar gaps due to the fact that satellite orbits used

J. Klokočník et al., *Gravitational Atlas of Antarctica*, Springer Geophysics,
DOI 10.1007/978-3-319-56639-9_3

for their computation are inclined to the Earth's equator with an angle different from 90°. GRACE has an orbit inclination of 89° and GOCE of 96.7°, so that the polar gap without the data coverage is as large as ~1° and ~7° around the poles for the GRACE and GOCE, respectively (Fig. 3.1). The decision on the inclination of a particular satellite mission depends on several factors, it is always a compromise between diverse scientific goals and also available technical means (e.g., Vallado 2007).

Because the GRACE data have a lower resolution than the GOCE data, then, going from the ocean to Antarctica, first we lose precision and resolution having only GOCE + GRACE data and no terrestrial data and then, very close to the poles, again we lose precision because we have only the GRACE data (Fig. 3.1). The narrow polar gap ~1° around the South Pole is without any data in EIGEN 6C4 and therefore, the aspects (2.2)–(2.7) and *vd* computed and plotted should be ignored in this region (see also Sect. 3.3).

Now both the Arctic and Antarctic polar gaps are nearly filled by airborne data from several sources in the frame of PolarGap project (e.g., Forsberg et al. 2016), but these results were not yet included nor into the EIGEN 6C4 model (see above), nor into the RET 14 (or similar model, see below). But one can expect activity of researches in this direction.

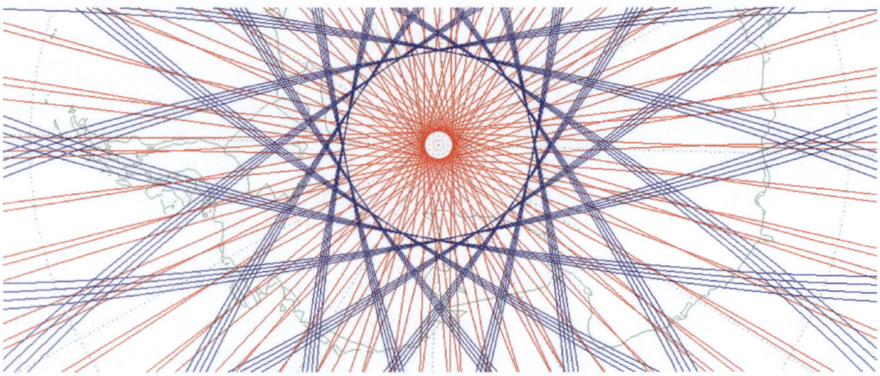

Fig. 3.1 An example of the ground tracks of satellite missions GRACE (*red*) and GOCE (*blue*) near the South Pole. Limiting latitudes of turning points between the ascending and descending tracks are given by the orbital inclination (the angle between the Earth's equator and the plane of the satellite orbit around the Earth), which is 89° for GRACE and 96.7° for GOCE. It means that not all latitudes are covered by satellite measurements. This explains the existence of the polar gaps without data, here about 1° and ~7° around the poles for the GRACE and GOCE, respectively. The ground track points of both missions for this example are displayed for the interval of 5 days in December 2010 (this interval is chosen randomly)

3.2 Bedrock Topography Model BEDMAP 2

There is a fantastic tool which enables one to see what is beneath the ground or ice (e.g., an ice–penetrating radar used in NASA's IceBridge airborne mission, www.nasa.gov/icebridge). The topography of the ground under the ice (bedrock, base of the ice sheets) in Antarctica, known as *Bedmap 2*, was compiled from 25 million of measurements of various kinds (satellite and air-borne) with a spatial resolution reaching 5 × 5 km (Fretwell et al. 2013), in a grid 1 × 1 km. But the resolution (and the quality of the bed topography) is still much worse for some areas of Antarctica without data (Fretwell et al. 2013, in Princess Elizabeth Land, in our Chap. 5 see the Segment 3 and Recovery and Support Force glaciers, in Chap. 5 here the Segments 11 and 12).

The data file ("bedmap2_bed") contains the bedrock elevation beneath the grounded ice sheet of Antarctica in the grid 1 × 1 km in the polar stereographic projection based on the WGS 84 ellipsoid. Errors in the ice sheet bed elevation are ~60 m in regions where there is the data, rising to 200 m in regions farther from observations (<20 km), and may be larger than 1000 m in regions more than 50 km away from any direct measurements! This is a bit risky for user because he/she does not know exactly quality of the data in the particular locality; but there is a rough guide, see Figs. 3 and 12 in Fretwell et al. (2013).

Bonvalot et al. (2012) combined topographic and gravity data to create maps, e.g., with Bouguer or isostatic gravity anomalies for Antarctica, but we need data in terms of C_{lm}, S_{lm}; we have now (since 2016) an excellent chance (Sect. 3.3).

3.3 A Combination of the Gravity and Bedrock Topography Data RET 14

A new degree-2190 gravity field model *SatGravRET2014* (shortly RET14) given as a set of the geopotential coefficients C_{lm}, S_{lm}, valid or meaningful from the physical viewpoint only for the continent of Antarctica, has been published by Hirt et al. (2016); it combines EIGEN 6C4 and Bedmap 2. It is obvious that RET14 increases the resolution of EIGEN 6C4 and decreases the resolution of Bedmap 2; generally its spatial resolution should be around 10 km (Hirt et al. 2016). This our Atlas is based on the RET14 model.

3.4 Other Data

Other important data sets have appeared recently, but not in terms of C_{lm}, S_{lm}, thus not directly compatible with our theory. Operation *IceBridge* was a 2009–2016 NASA mission aiming to monitor changes in polar ice from a fixed-wing aircraft (e.g., Studinger and Martin 2010). It is a temporary replacement for the satellite ICESat until the ICESat-2 launch expected in 2017. The new Antarctic gravity anomaly grid (Scheinert et al. 2016) incorporates all available gravity data over Antarctica over the last three decades. The PolarGap project (Forsberg et al. 2016) nearly filled the GOCE polar gap (Sect. 3.1) with the airborne gravity, magnetic and radar data, reaching precision 1–2 mGal rms in Δg. A part of the IceBridge data entered Bedmap 2 and so the RET14 too, "indirectly".

3.5 Practical Guide to Figures in this Atlas

Plotting figures—nonlinear scales. In our figures of the gravitational aspects and topography, non-linear color scales are used to emphasize various features which otherwise might remain hidden or overlooked. The gravity anomalies and/or disturbances are in milliGals [mGal], the second order potential derivatives are in Eötvös; recall that 1 mGal = 10^{-5} ms^{-2}, 1 E = 1 Eötvös = 10^{-9} s^{-2}. The invariants I_1 and I_2 have units [s^{-4}] and [s^{-6}] and the ratio I is dimensionless. The strike angle θ is expressed in degrees with respect to the local meridian and its display in red means its direction to the east and in blue to the west of the meridian. These units are used in all presented figures.

Projection. Our figures for Antarctica are in stereographic polar projection with a non-deformable parallel at 71° South (compare with Fretwell et al. 2013, p. 377). This is because in this projection we obtained the topographic model Bedmap 2. The rectangular plain coordinates X, Y in the stereographic polar projection were computed from the known geocentric latitude ϕ and longitude λ using these equations:

$$r = r_e.(1 + \sin 71^\circ).\tan[(90^\circ - |\phi|)/2]$$
$$X = r.\sin\lambda, Y = r.\cos\lambda$$

where r_e = 6378 km is the radius of a reference sphere for the Earth.

The main series of the figures (next section) looks like a slice of a cake; we call them "**segments**". Each segment contains all the gravitational aspects plus topography. The segments go from the southern latitude 60° to the South Pole (−90°). Thus the polar gap area (Sect. 3.1) shown in our maps should be ignored; the results in the vicinity of the South Pole (90 ± 1°) are plotted but they are unrealistic. In longitude (E…counted East of Greenwich), the segments contain 30° intervals plus 5° overlaps; it means the first segment goes from 0 to 35°E, the second from 30° to 65°E, the third from 60° to 95°E, till 330°–365° (= 5°E).

Virtual deformations of the ellipse of deformation (*vd*) are geometrically expressed by its **dilatation** (shown in red color) or **compression** (blue). The plotted quantities are a and b, see the list below Eq. 2.12, expressed in our figures as crosses. The virtual dilatation indicates uplifted regions at the geoid, whose mass has a tendency to disintegrate. On the contrary, the virtual compression indicates lowered zones at the geoid.

Natural processes, which are the cause of these features near the surface of the geoid, are certainly very diverse as a consequence of the regionally heterogeneous integration of morphotectonic, erosion-denudation processes, and others. But it is NOT the purpose of this Atlas to interpret; the interpretations are on geologists, geophysicists, glaciologists, climatologists, morphologists, oceanologists, geodesists, global change people, and others.

References

Bonvalot S., Balmino G., Brias A., Kuhn M., Peyrefitte A. and Vales N. 2012. World Gravity Map. Bureau Gravimetr. Interntl., map, edited by CGMW-BGI-CNES-IRD, Paris.

Floberghagen, R., Fehringer, M., Lamarre, D., et al. 2011. Mission design, operation and exploitation of the gravity Field and steady-state ocean circulation explorer mission, *J. Geod.* 85: 749–758; http://dx.doi.org/10.1007/s00190-012-0541-z.

Fretwell P, Pritchard H, Vaughan D, Bamber J, et al. 2013. Bedmap2: improved ice bed, surface and thickness datasets for Antarctica. *The Cryosphere*, 7, 375–393, doi:10.5194/tc-7-375-2013.

Förste (Foerste) Ch., Bruinsma S., Abrykosov O., Lemoine J-M. et al. 2014. The latest combined global gravity field model including GOCE data up to degree and order 2190 of GFZ Potsdam and GRGS Toulouse (EIGEN 6C4). *5th GOCE user workshop*, Paris 25–28, Nov.

Forsberg, R., Olesen, A.V., Ferraccioli, F., Jordan, T., Matsuoka, K., Kohler, J. 2016. GOCE polar gaps now filled – first gravity results of the ESA PolarGap project, presented at the *ESA Living planet Symp.*, Prague, Czech Republic.

Hirt Ch., Rexer M., Scheinert M., Pail R., Claessens S. and S. Holmes 2016. A new degree-2190 (10 km resolution) gravity field model for Antarctica developed from GRACE, GOCE and Bedmap 2 data, *J. Geod.* 90: 105–127; doi:10:1007/s00190-015-0857-6.

Pavlis N. K., Holmes S. A., Kenyon S. C. and Factor, J. K. 2008a. An Earth Gravitational Model to degree 2160: EGM 2008. *EGU General Assembly* 2008, Vienna, 13–18 April.

Pavlis N. K., Holmes S. A., Kenyon S. C. and Factor, J. K. 2008b. EGM2008: An overview of its development and evaluation. National Geospatial-Intelligence Agency, USA, conf.: *Gravity, Geoid and Earth Observation 2008*, Chania, Crete, Greece, 23–27 June.

Pavlis N. K., Holmes S. A., Kenyon S. C. and Factor J. K. 2012. The development and evaluation of the Earth Gravitational Model 2008 (EGM2008). *J. Geophys. Res.* 17: B04406, doi:10.1029/2011JB008916, 2012.

Scheinert M., Ferraccioli, F., Schwabe, J., Bell, R. E., et al., T. 2016. New Antarctic gravity anomaly grid for enhanced geodetic and geophysical studies in Antarctica. *Geophys. Res. Letts*, doi:10.1002/2015GL067439.

Studinger, M. L., Martin, S., Sonntag J., 2010. Operation Icebridge, presented at *IEEE conf.*, pp. 1918–1919, Honolulu, Hawaii, 25-30 July; doi:10.1109/ IGARSS.2010.5650555.

Tapley, B. D., Bettadpur, S., Watkins, M., and Reigber, C. 2004. The gravity recovery and climate experiment: Mission overview and early results, *Geophys. Res. Letts*, 31: 9607.

Vallado, D.A., 2007. Fundamentals of Astrodynamics and Applications. Springer.

Chapter 4
Antarctica—A General Look

The quantities (2.2)–(2.7) and *vd* were computed and plotted for the whole planet. "General views" over the whole Antarctica are presented in this section (the same as in Klokocník et al. 2016, but all with RET 14). The topography of bedrock from Bedmap 2 is added (including the names of the main topographic structures), see the series of Figs. 4.1, 4.2, 4.3, 4.4, 4.5, 4.6, 4.7, 4.8 and 4.9.

The reader is invited to ask the authors of this Atlas to provide him/her with high resolution PNG files containing Figs. 4.1, 4.2, 4.3, 4.4, 4.5, 4.6 and 4.7; they can be printed in a large format as posters.

It is evident that these figures cannot provide too much detail. Therefore, more details are in the next chapter—Chap. 5. It contains the essential part of this Atlas. Antarctica is divided into the segments mentioned above, separated by geographic longitude, where all the gravitational aspects are depicted. There are also additional "zooms" in the individual segments showing selected interesting locations. On request further examples can be prepared.

The values of *vd* over Antarctica are in Fig. 4.7 with a reduced resolution to be able to plot this comprehensive figure (for more details, see below the figures with *vd* depicted for the individual segments and localities). The areas with dilatation are in the red colour, those of compression in blue. The strongest stresses appear to follow the continental-ocean transition, and elevated topography of the Transantarctic Mountains. They are located also in a belt going along longitude 160° roughly from the south to the vicinity of the Pole, at the Antarctic Peninsula (west), and near the South Pole.

J. Klokočník et al., *Gravitational Atlas of Antarctica*, Springer Geophysics,
DOI 10.1007/978-3-319-56639-9_4

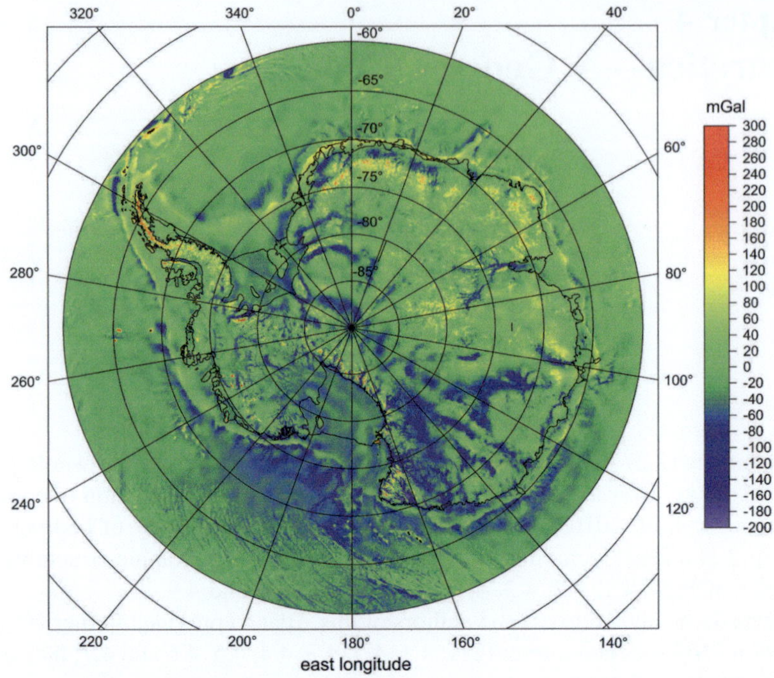

Fig. 4.1 The values of Δg [mGal] over Antarctica

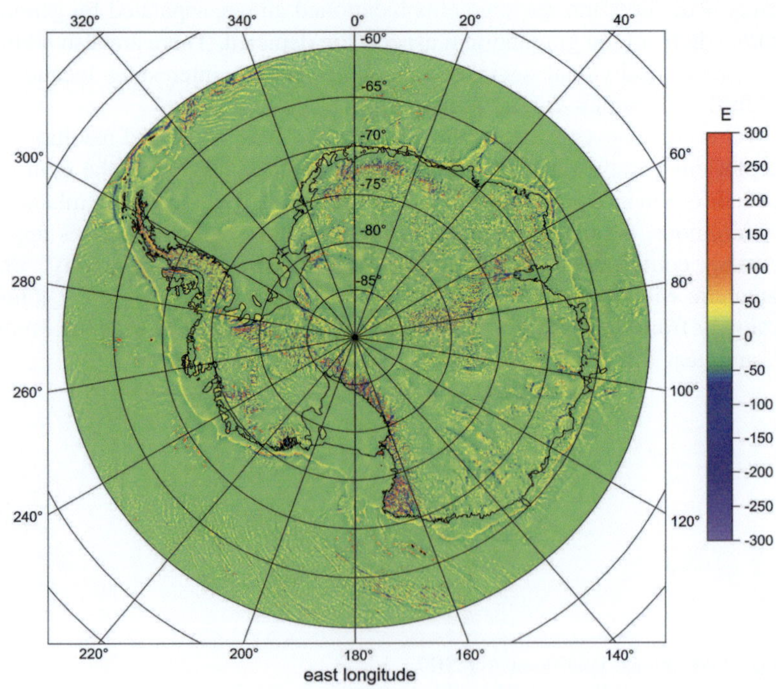

Fig. 4.2 The values of $T_{zz,}$ scale in [Eötvös], over Antarctica

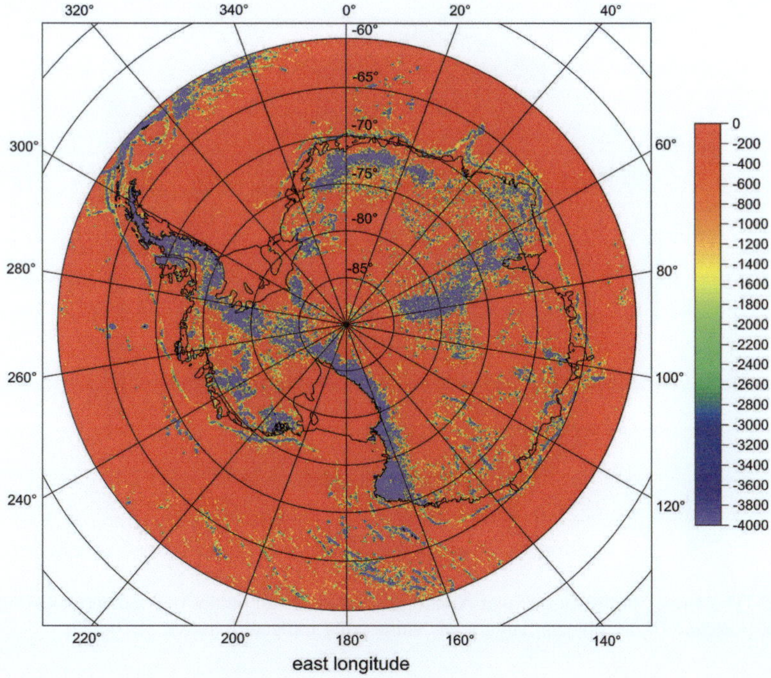

Fig. 4.3 The values of the invariant I_1 [s^{-4}] over Antarctica

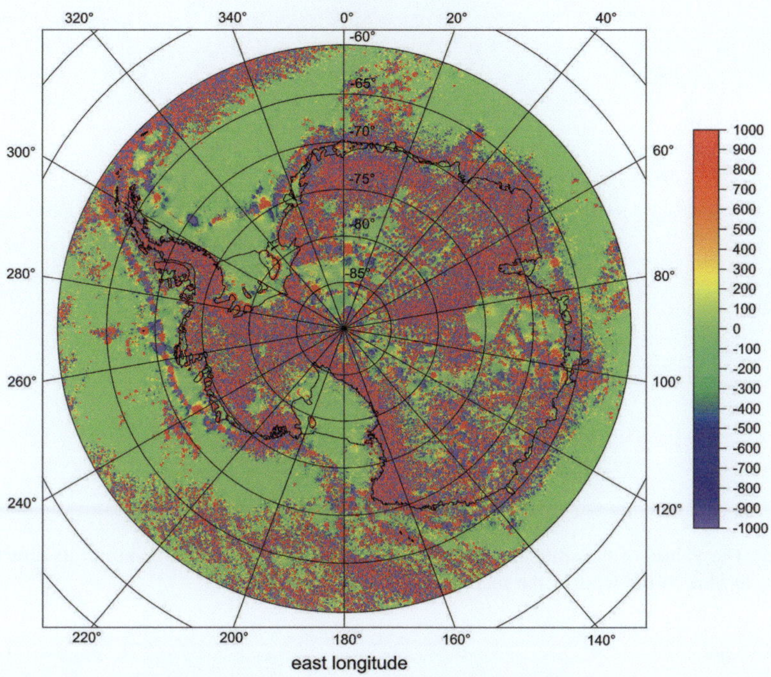

Fig. 4.4 The values of the invariant I_2 [s^{-6}] over Antarctica

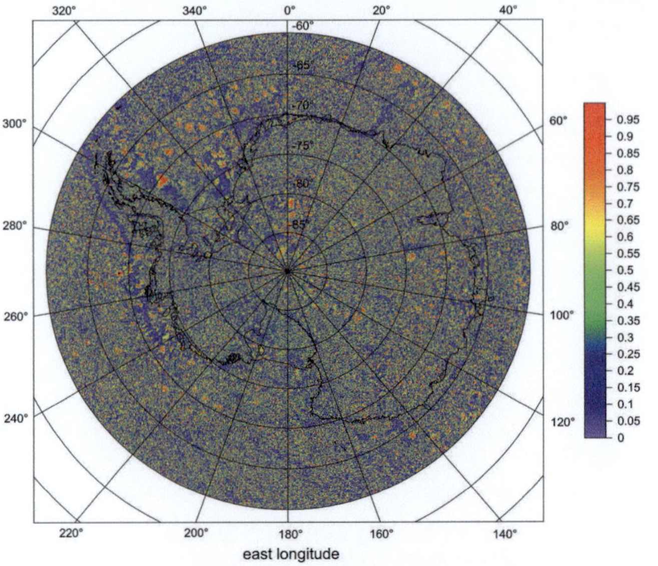

Fig. 4.5 The values of the ratio I over Antarctica. Small (*blue*) values of I indicate localities with possible presence of flat objects, large (*red*) values of I indicate massive 3D bodies

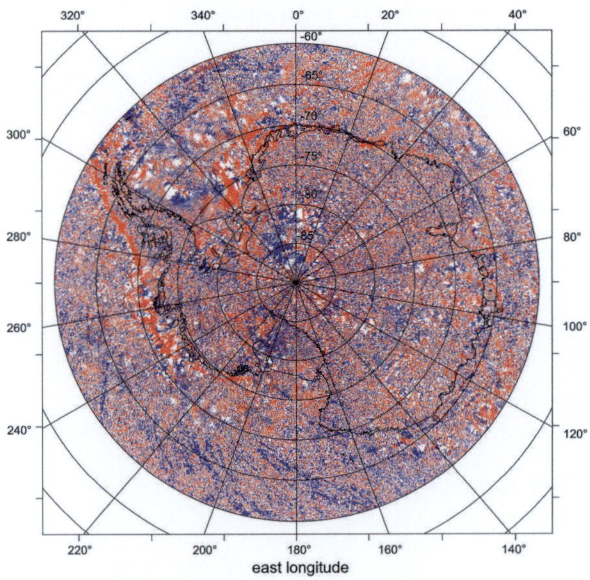

Fig. 4.6 The values of the strike angle θ over Antarctica (scale in degrees: in *red* its direction to the east, in *blue* to the west of the meridian)

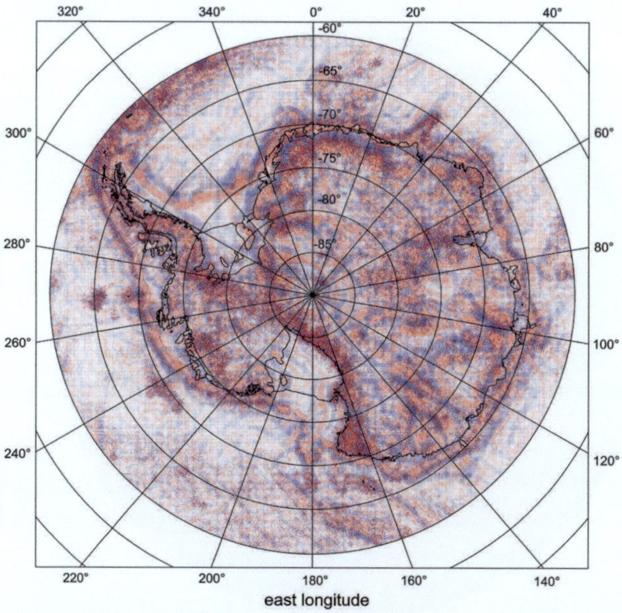

Fig. 4.7 The values of *vd* over Antarctica (*red* for the dilatation, *blue* for the compression)

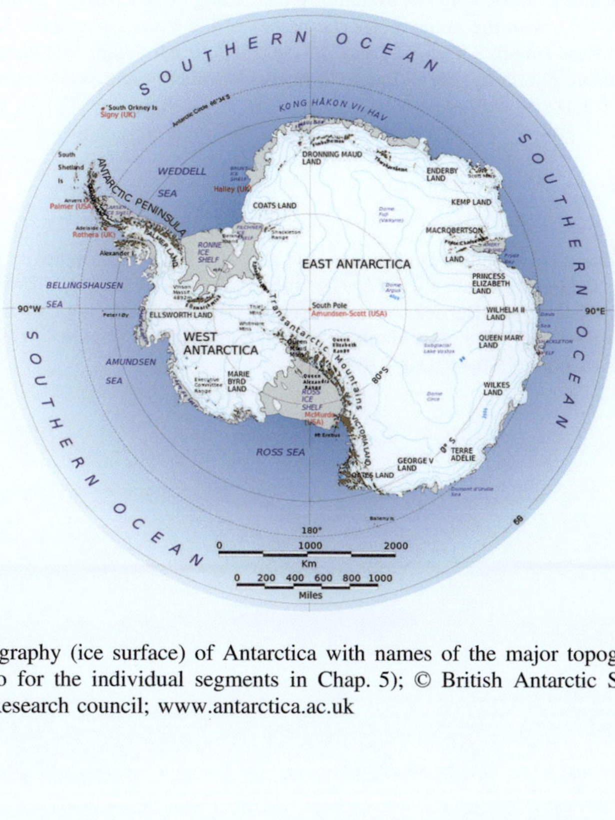

Fig. 4.8 Topography (ice surface) of Antarctica with names of the major topographic features (used later also for the individual segments in Chap. 5); © British Antarctic Survey, Natural Environment Research council; www.antarctica.ac.uk

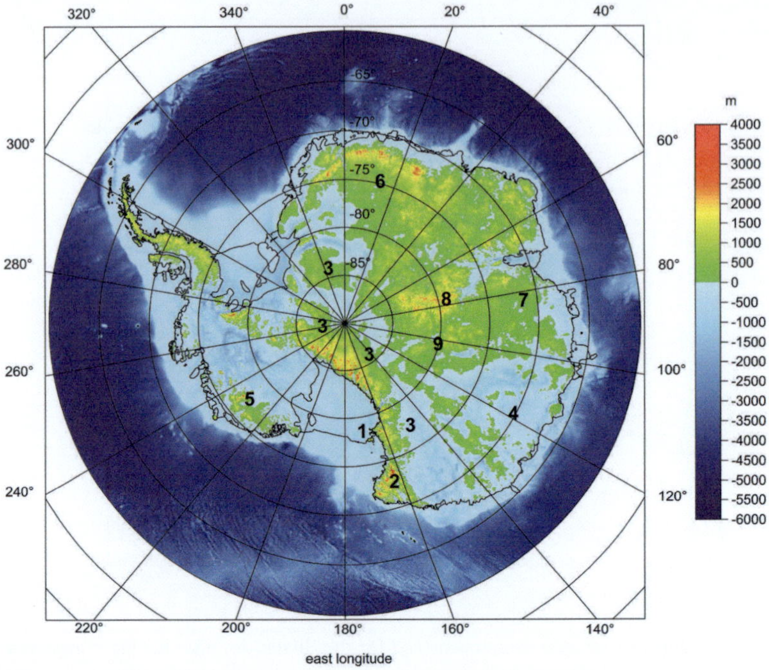

Fig. 4.9 The bedrock topography in Antarctica (according to BEDMAP 2) in grid 1×1 km (Fretwell et al. 2013), with the numbers for the names of the main topographic features: *1* Ross shelf glacier and Ross Island, *2* Victoria Land. *3* Transantarctic Mountain. *4* Wilkes Land. *5* Marie Byrd Land. *6* Queen (Dronning) Maud Land. *7* Princess Elizabeth Land. *8* Gamburtsev Mountains (under the ice). *9* Lake Vostok area

References

Fretwell P., Pritchard H., Vaughan D., Bamber J., et al. 2013. Bedmap2: improved ice bed, surface and thickness datasets for Antarctica. *The Cryosphere*, 7, 375–393, doi: 10.5194/tc-7-375-2013.

Klokočník J., Kostelecký J., Bezděk A. and Pešek I. 2016. On feasibility to detect volcanoes hidden under the ice of Antarctica via their "gravitational signal", *Annals of Geophysics* 59(5), S0539. http://dx.doi.org/10.4401/ag-7102.

Chapter 5
Segments and Selected Details

The figures are organized in **segments**. Each segment starts with an "introductory" figure (looking like "a cut or a piece of cake"), accompanied with "wind rose", showing its latitudinal and longitudinal range and its position in Antarctica (a little "balloon"). These figures are labelled **Fig. 5.i.0**. Each segment contains all the gravitational aspects Δg, T_{zz}, I_1, I_2, I, θ (j = 1, 2, 3, …. 6), then *vd*, finally the topography from BEDMAP 2, in total ten like-a-cake-cut figures per segment (10 because for the virtual deformations *vd*, we present their total value and then separately their dilatational and compressional parts). The figure numbers have a format given as **Fig. 5.i.j**, ordered first by segment **i** and then, inside the given segment, by the gravitational aspect **j**. The first segment covers longitudes from 0 to 35°E, the second from 30 to 65°E, etc., till the last segment which is from 330 to 365°E, so we have twelve segments together. The range of geodetic latitude of each segment always goes from −60° to −90°. *Examples*: Fig. 5.2.2 shows T_{zz} for the segment with longitudes between 30°E and 65°E, Fig. 5.10.6 shows θ for the segment with longitudes between 270°E and 305°E. We may summarize:

$$\textbf{Fig. 5.i.j}, \quad \text{where} \quad i = 1{-}12,\ j = 1{-}10.$$

Each segment starts with **Fig. 5.i.0**, which shows the wind rose and the scheme ("introductory figure"). For the majority of the segments we add **zooms** (close-ups, local views) with various details based on BEDMAP 2 or RET 14, because we guess that such examples might be interesting for geoscientists. These are mountain ranges, valleys, faults, nunataks (rocks protruding through the ice), subglacial lakes, hypothetical volcanoes hidden under the ice, etc. We are very far from a systematic

J. Klokočník et al., *Gravitational Atlas of Antarctica*, Springer Geophysics,
DOI 10.1007/978-3-319-56639-9_5

detailed scanning of the whole Antarctica with all the gravitational aspects in this Atlas; these zooms are only examples showing what we are able to generate with the existing data and software. The term "detailed" in our scale means "with resolution of about ten kilometres" (we recall the notes about the resolution of the RET 14, ~10 km, Sect. 3.3).

The reader is invited to contact the authors of the Atlas to receive more examples for the areas of his/her interest (jklokocn@asu.cas.cz).

See Figs. 5.1.0, 5.1.1, 5.1.2, 5.1.3, 5.1.4, 5.1.5, 5.1.6, 5.1.7, 5.1.8, 5.1.9, 5.1.10, 5.1.11, 5.1.12, 5.1.13, 5.1.14 and 5.1.15.

See Figs. 5.2.0, 5.2.1, 5.2.2, 5.2.3, 5.2.4, 5.2.5, 5.2.6, 5.2.7, 5.2.8, 5.2.9, 5.2.10, 5.2.11, 5.2.12 and 5.2.13.

See Figs. 5.3.0, 5.3.1, 5.3.2, 5.3.3, 5.3.4, 5.3.5, 5.3.6, 5.3.7, 5.3.8, 5.3.9, 5.3.10, 5.3.11, 5.3.12, 5.3.13, 5.3.14, 5.3.15, 5.3.16, 5.3.17 and 5.3.18.

See Figs. 5.4.0, 5.4.1, 5.4.2, 5.4.3, 5.4.4, 5.4.5, 5.4.6, 5.4.7, 5.4.8, 5.4.9, 5.4.10, 5.4.11, 5.4.12, 5.4.13 and 5.4.14.

See Figs. 5.5.0, 5.5.1, 5.5.2, 5.5.3, 5.5.4, 5.5.5, 5.5.6, 5.5.7, 5.5.8, 5.5.9, 5.5.10, 5.5.11, 5.5.12, 5.5.13 and 5.5.14.

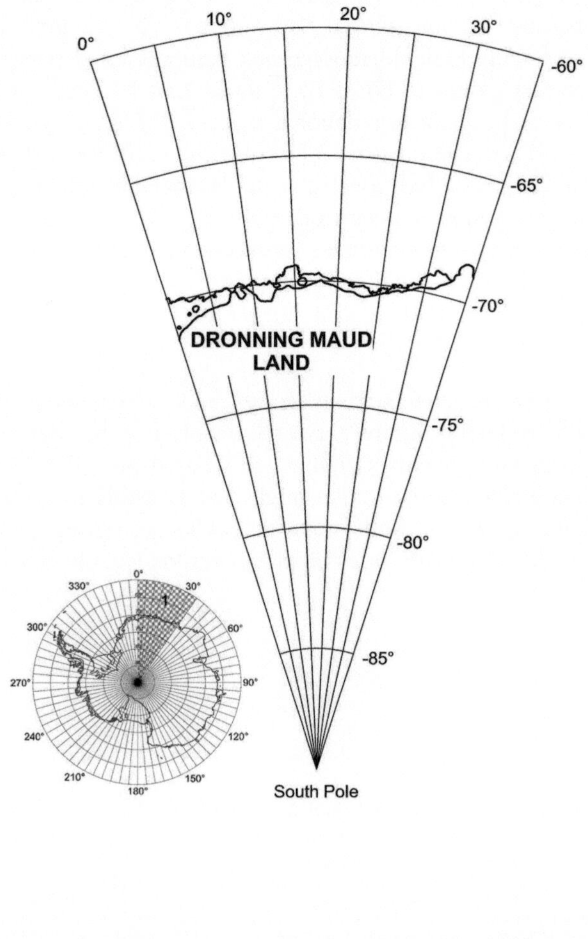

Fig. 5.1.0 An orientation rose for this segment (inset, balloon, wind rose) and the introductory figure for this segment showing its coordinate network

Fig. 5.1.1 Δg [mGal]

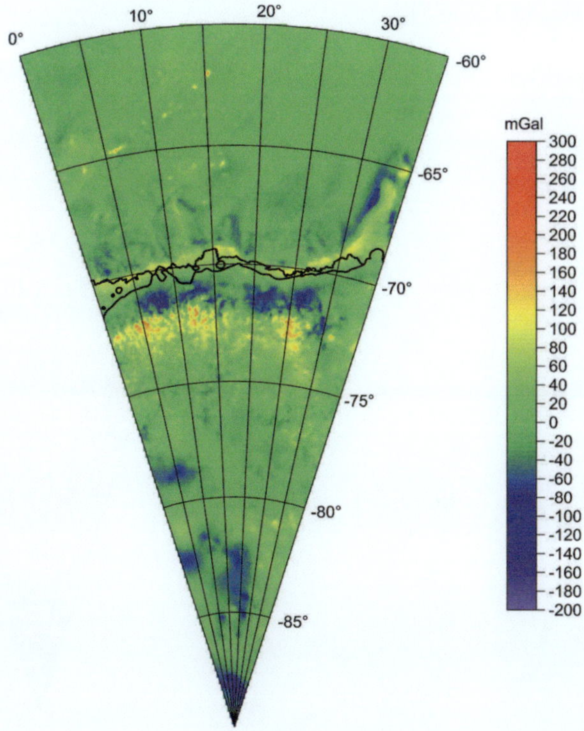

Fig. 5.1.2 T_{zz} [E]

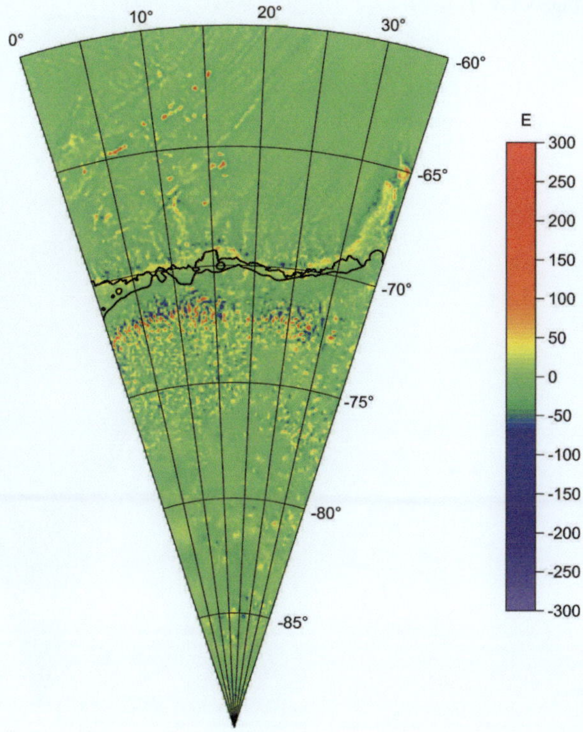

Fig. 5.1.3 I_1 [s^{-4}]

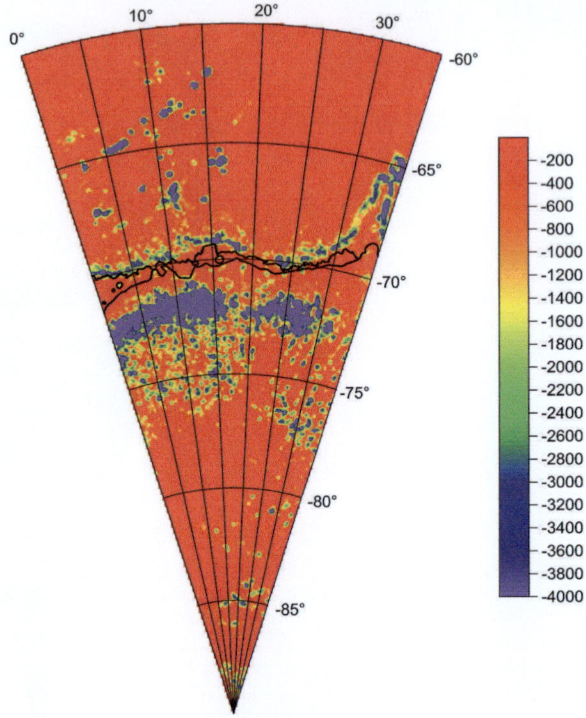

Fig. 5.1.4 I_2 [s^{-6}]

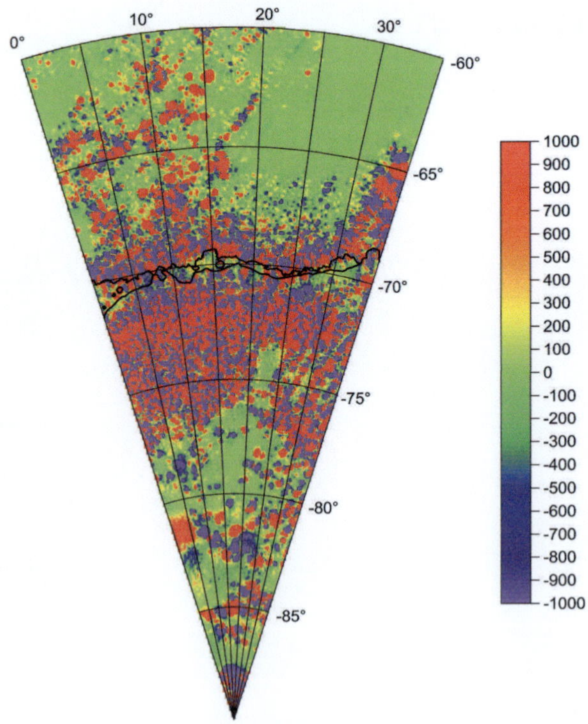

Fig. 5.1.5 I [−]

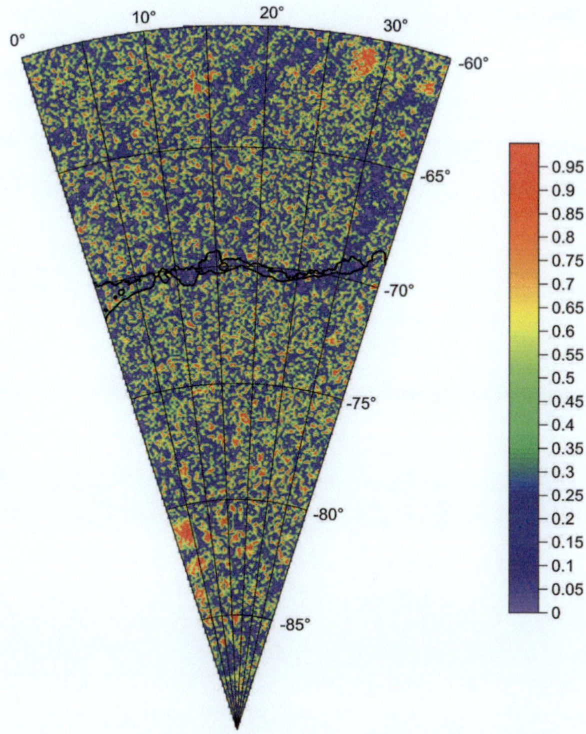

Fig. 5.1.6 θ

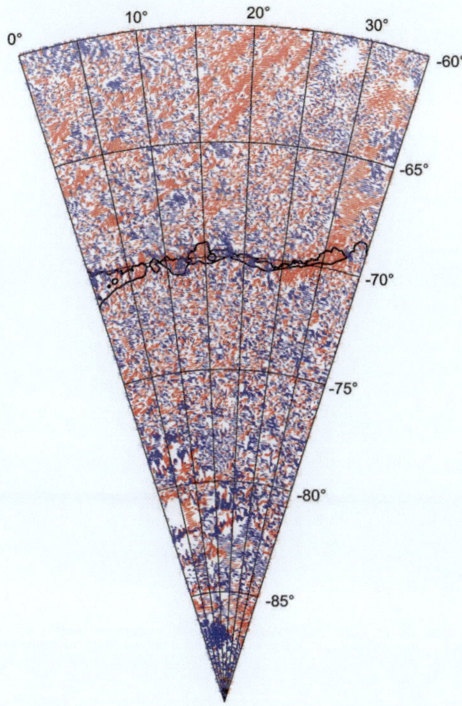

Fig. 5.1.7 *vd*

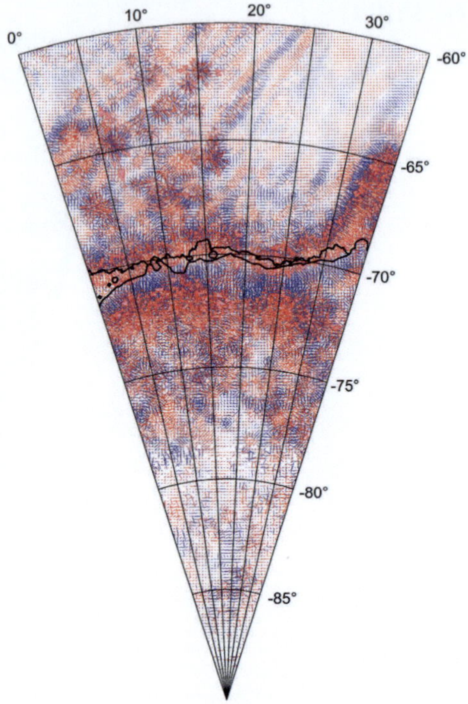

Fig. 5.1.8 *vd,* dilatation

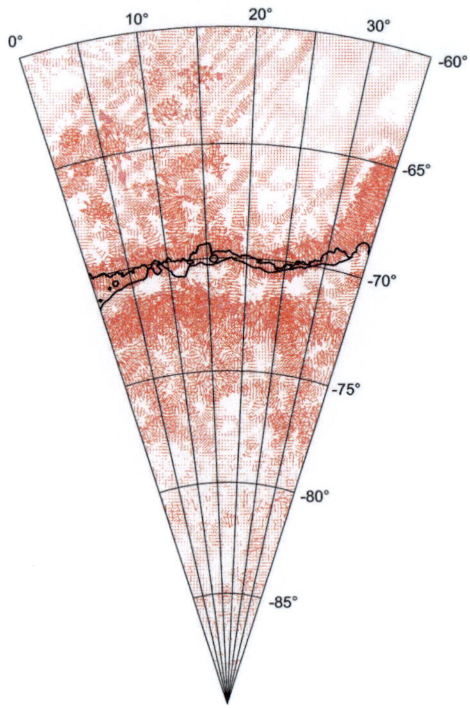

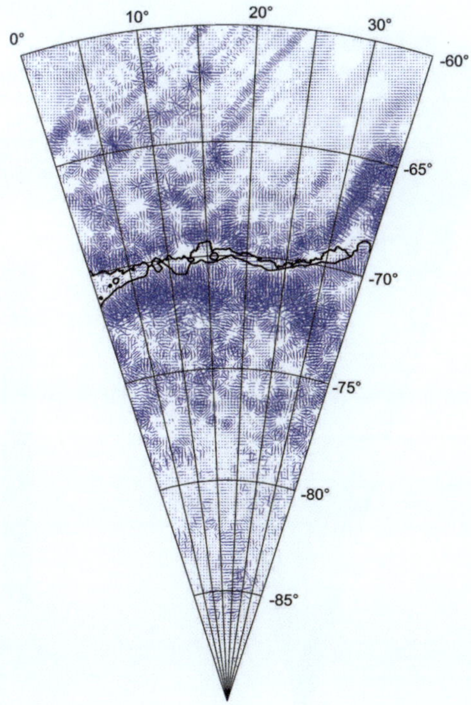

Fig. 5.1.9 *vd,* compression

See Figs. 5.6.0, 5.6.1, 5.6.2, 5.6.3, 5.6.4, 5.6.5, 5.6.6, 5.6.7, 5.6.8, 5.6.9, 5.6.10, 5.6.11, 5.6.12, 5.6.13, 5.6.14 and 5.6.15.

See Figs. 5.7.0, 5.7.1, 5.7.2, 5.7.3, 5.7.4, 5.7.5, 5.7.6, 5.7.7, 5.7.8, 5.7.9 and 5.7.10.

See Figs. 5.8.0, 5.8.1, 5.8.2, 5.8.3, 5.8.4, 5.8.5, 5.8.6, 5.8.7, 5.8.8, 5.8.9, 5.8.10, 5.8.11, 5.8.12, 5.8.13, 5.8.14, 5.8.15 and 5.8.16.

See Figs. 5.9.0, 5.9.1, 5.9.2, 5.9.3, 5.9.4, 5.9.5, 5.9.6, 5.9.7, 5.9.8, 5.9.9, 5.9.10, 5.9.11, 5.9.12, 5.9.13, 5.9.14, 5.9.15 and 5.9.16.

See Figs. 5.10.0, 5.10.1, 5.10.2, 5.10.3, 5.10.4, 5.10.5, 5.10.6, 5.10.7, 5.10.8, 5.10.9, 5.10.10, 5.10.11 and 5.10.12.

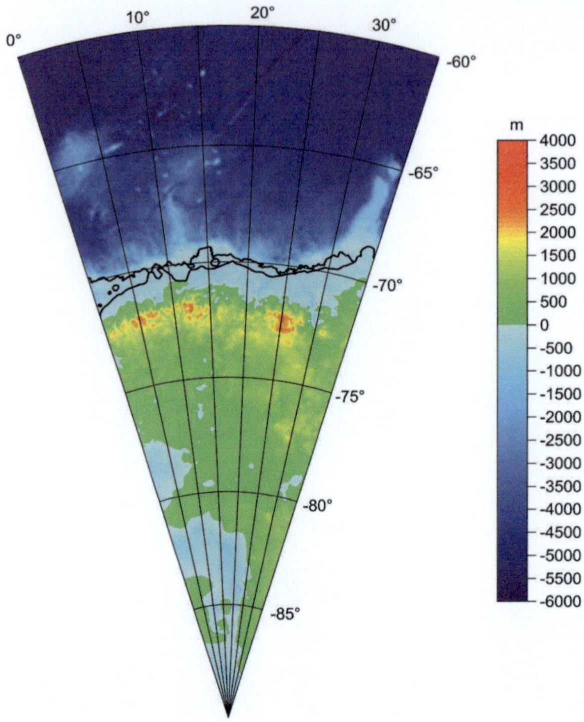

Fig. 5.1.10 Bedmap 2 topography [metres above sea level]

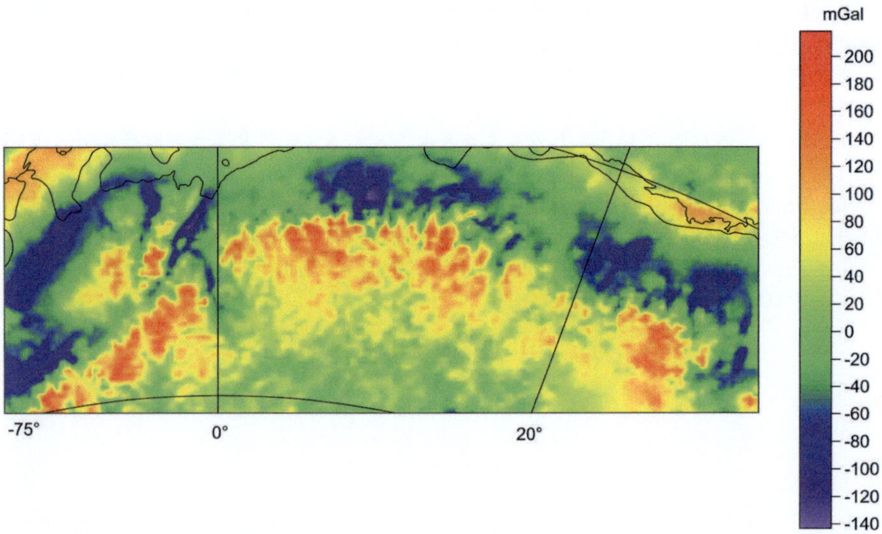

Fig. 5.1.11 Gravity disturbances Δg in a part of Dronning (Queen) Maud Land near the seashore. Topographic details of this area are shown in Figs. 5.1.13–5.1.15

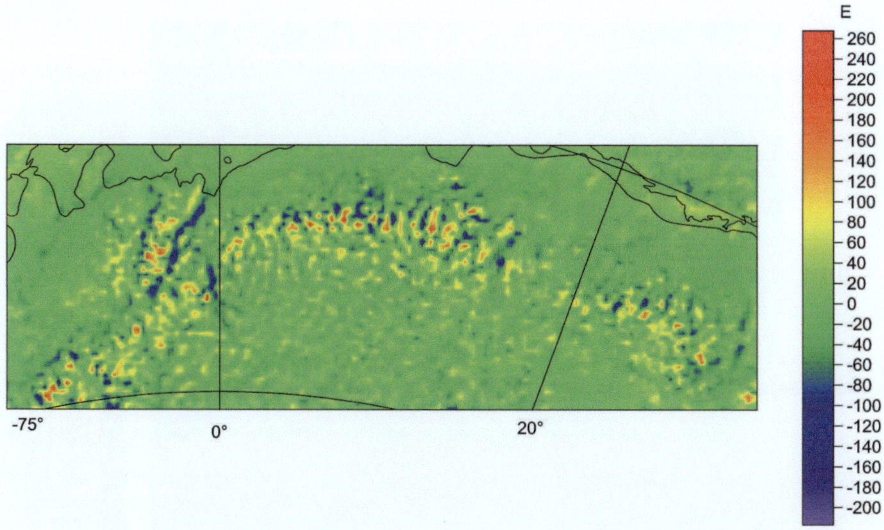

Fig. 5.1.12 The radial component of the Marussi tensor T_{zz} in a part of Dronning (Queen) Maud Land near the seashore. Topographic details of this area are shown in Figs. 5.1.13–5.1.15

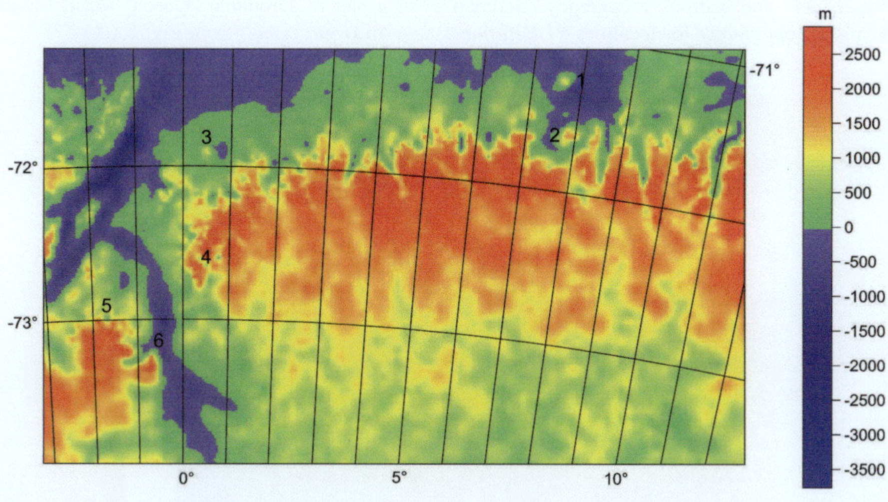

Fig. 5.1.13 The bedrock topography (Bedmap 2) in a part of Dronning (Queen) Maud Land and examples of the details of topography for localities #1 and #6 from Fig. 5.1.13

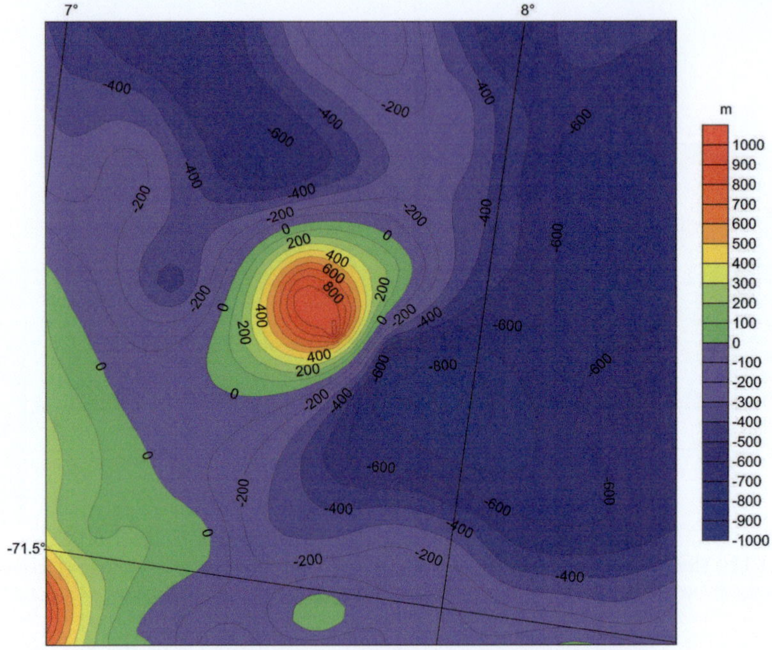

Fig. 5.1.14 The bedrock topography (Bedmap 2) in a part of Dronning (Queen) Maud Land: details of topography for localities #1 (*left panel*) and #6 (*right panel*) from Fig. 5.1.13

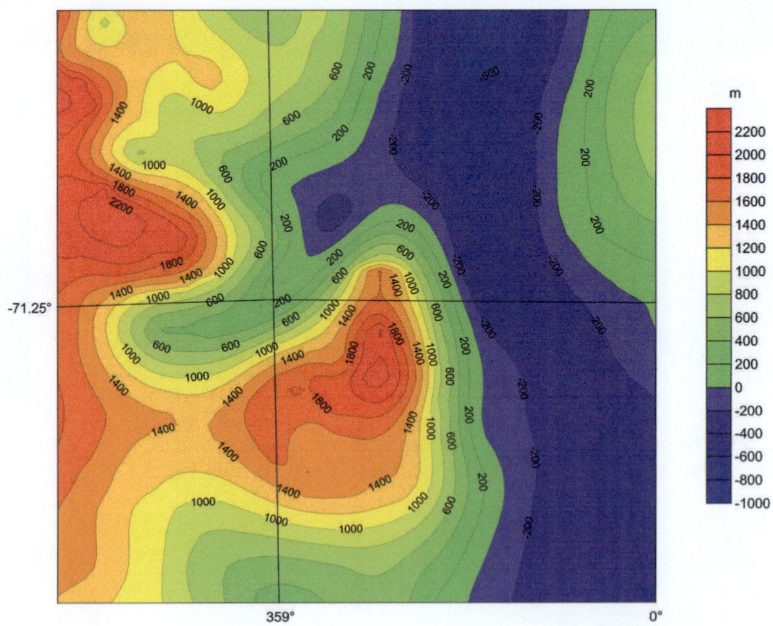

Fig. 5.1.15 The bedrock topography (Bedmap 2) in a part of Dronning (Queen) Maud Land: details of topography for localities #1 (*left panel*) and #6 (*right panel*) from Fig. 5.1.13

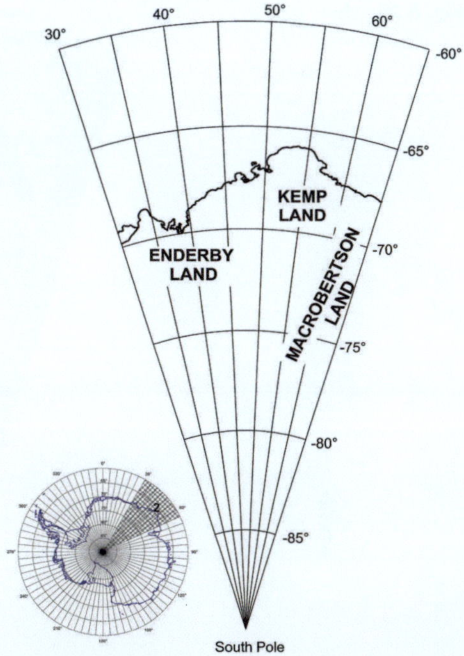

Fig. 5.2.0 An orientation rose for this segment (small figure) and introductory figure for this segment showing a few topographic features of the area

See Figs. 5.11.0, 5.11.1, 5.11.2, 5.11.3, 5.11.4, 5.11.5, 5.11.6, 5.11.7, 5.11.8, 5.11.9, 5.11.10, 5.11.11 and 5.11.12.

See Figs. 5.12.0, 5.12.1, 5.12.2, 5.12.3, 5.12.4, 5.12.5, 5.12.6, 5.12.7, 5.12.8, 5.12.9 and 5.12.10.

Fig. 5.2.1 Δg

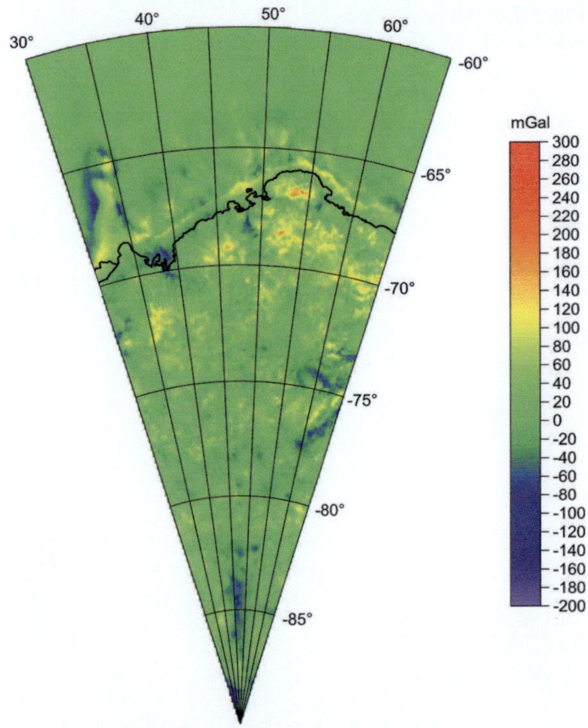

Fig. 5.2.2 T_{zz}

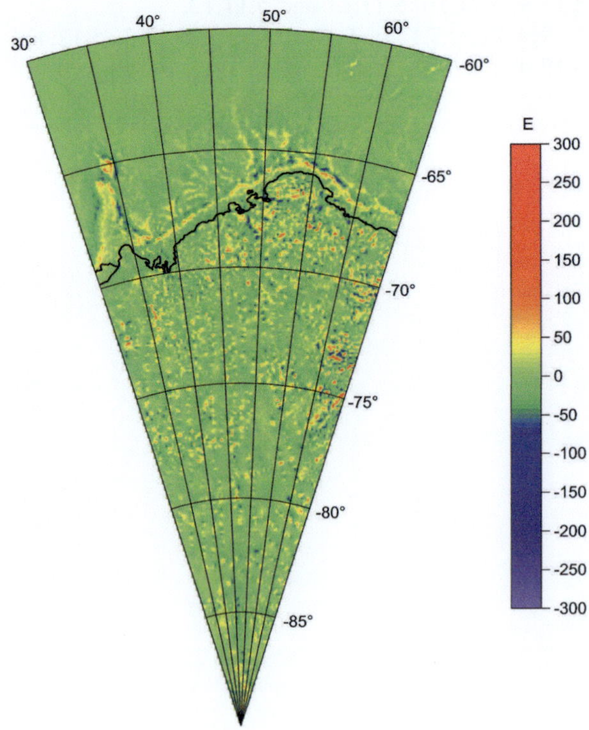

Fig. 5.2.3 I_1

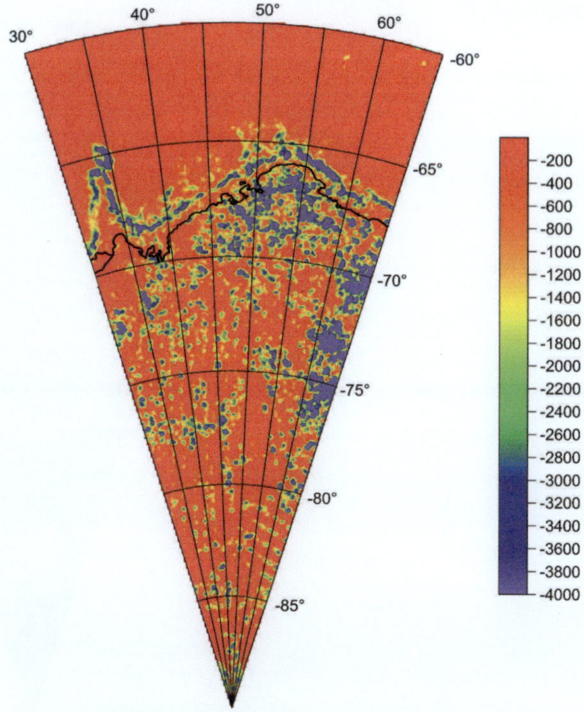

Fig. 5.2.4 I_2

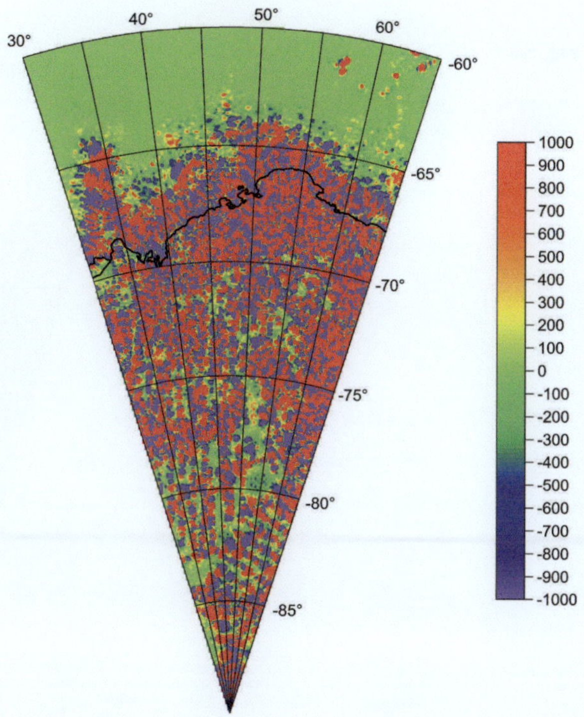

Fig. 5.2.5 I

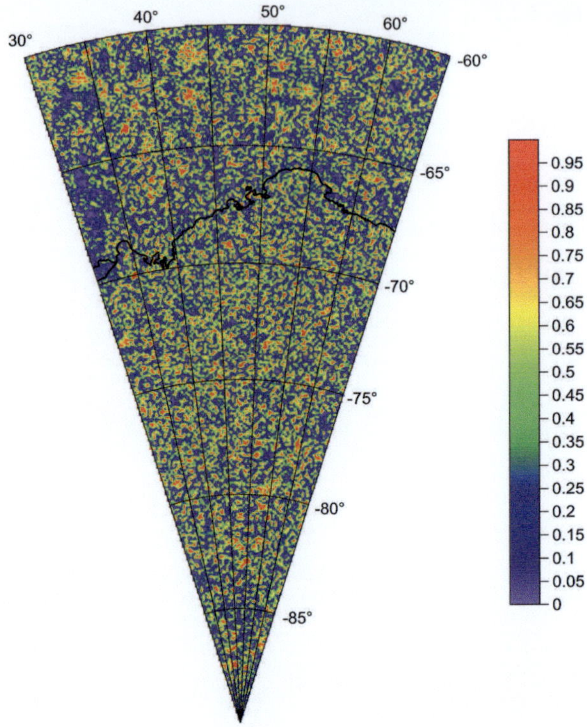

Fig. 5.2.6 θ

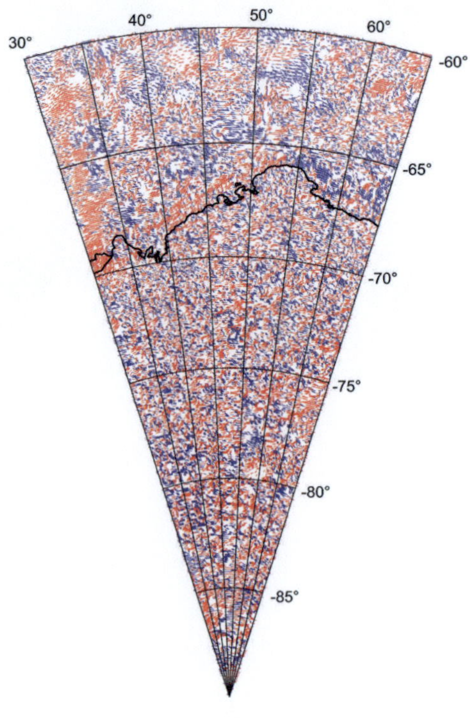

Fig. 5.2.7 *vd*

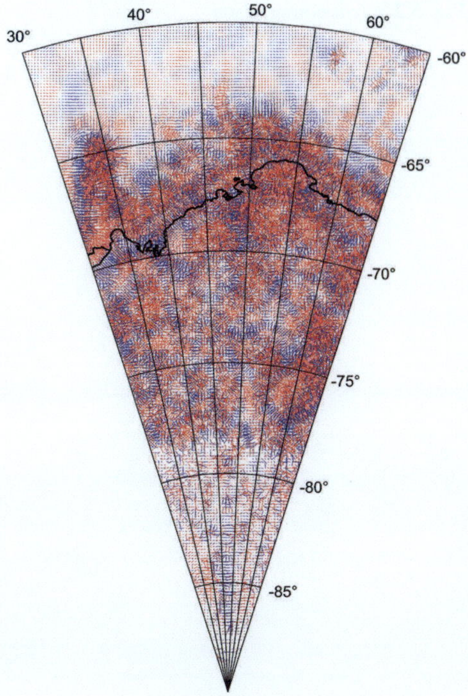

Fig. 5.2.8 *vd,* dilatation

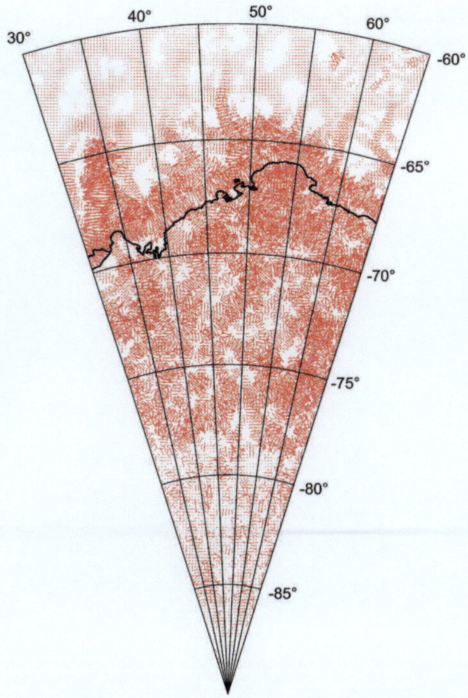

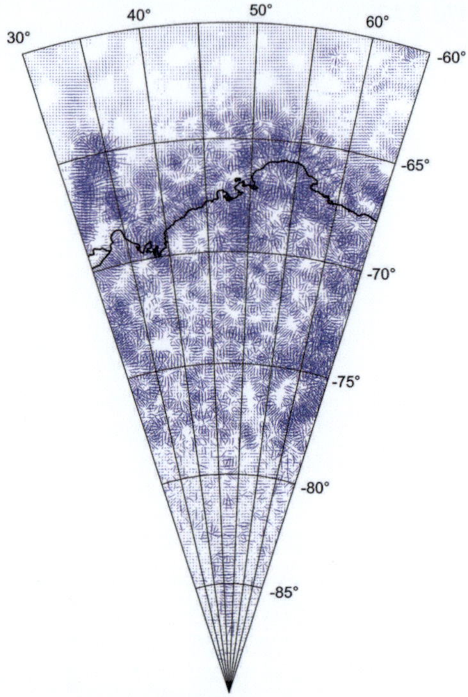

Fig. 5.2.9 *vd,* compression

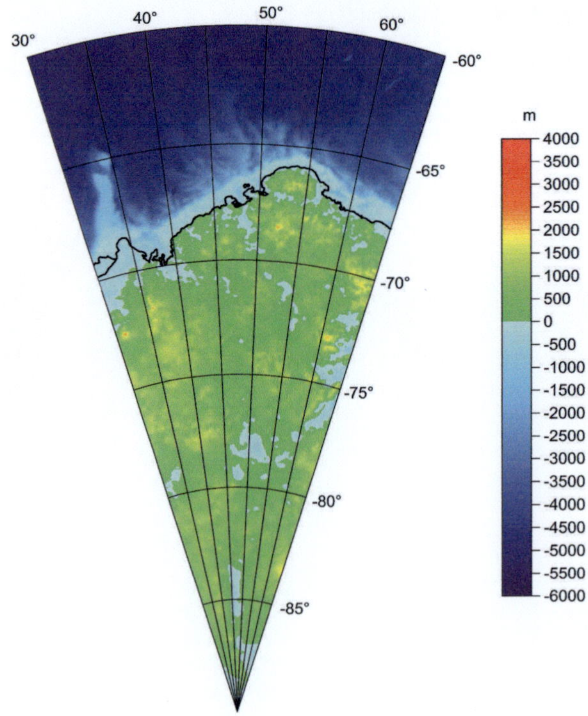

Fig. 5.2.10 Bedmap 2 topography

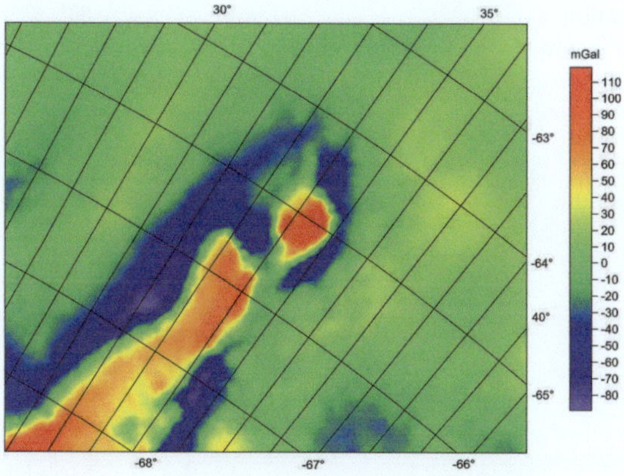

Fig. 5.2.11 A close-up of the sea area, Gunnerus Ridge, at $\varphi = -65°$, $\lambda = 35°E$, Δg

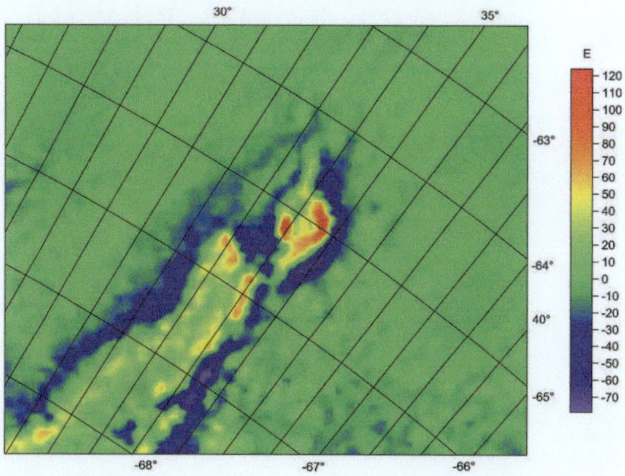

Fig. 5.2.12 A close-up of the sea area at $\varphi = -65°$, $\lambda = 35°E$, T_{zz}. The features on Figs. 5.2.11–5.2.13 are below today's sea level

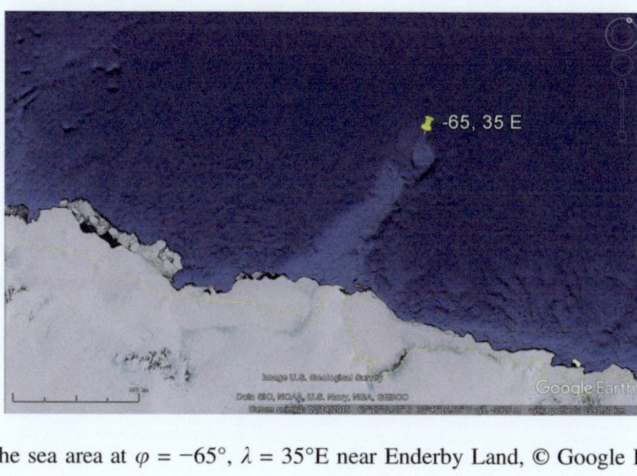

Fig. 5.2.13 The sea area at $\varphi = -65°$, $\lambda = 35°E$ near Enderby Land, © Google Earth

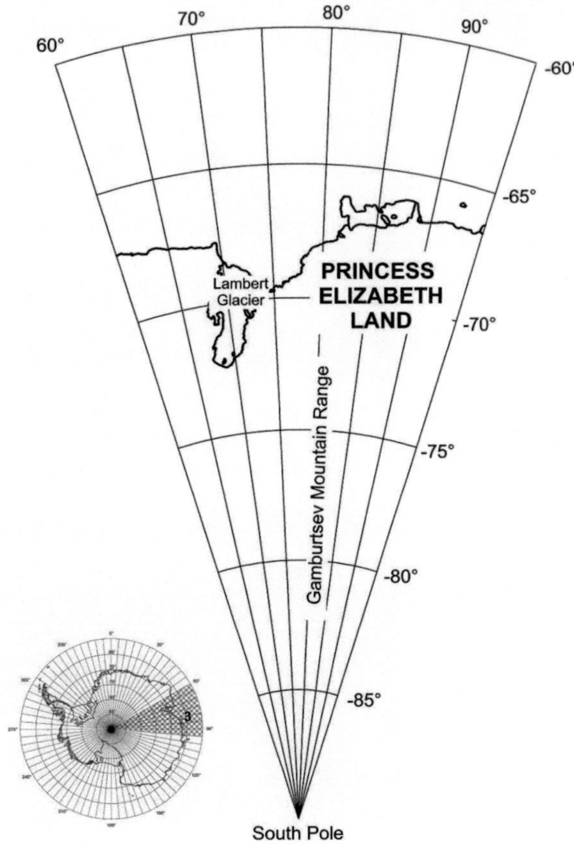

Fig. 5.3.0 An orientation rose and introductory figure for this segment

Fig. 5.3.1 Δg

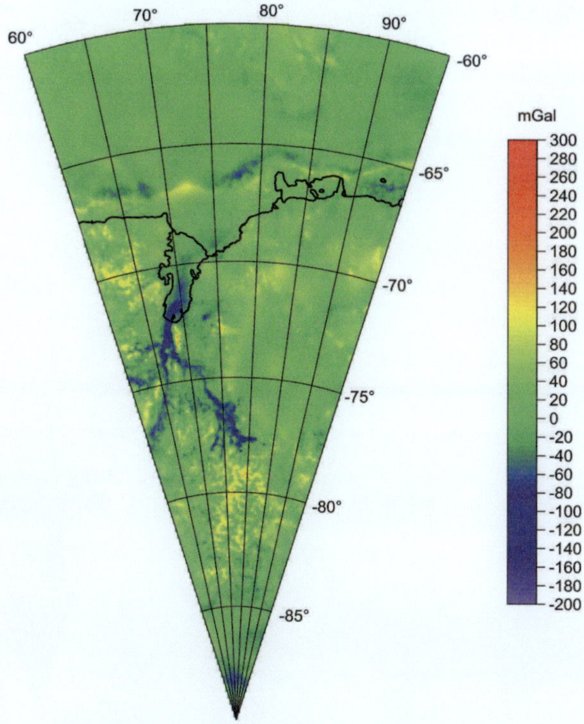

Fig. 5.3.2 T_{zz}

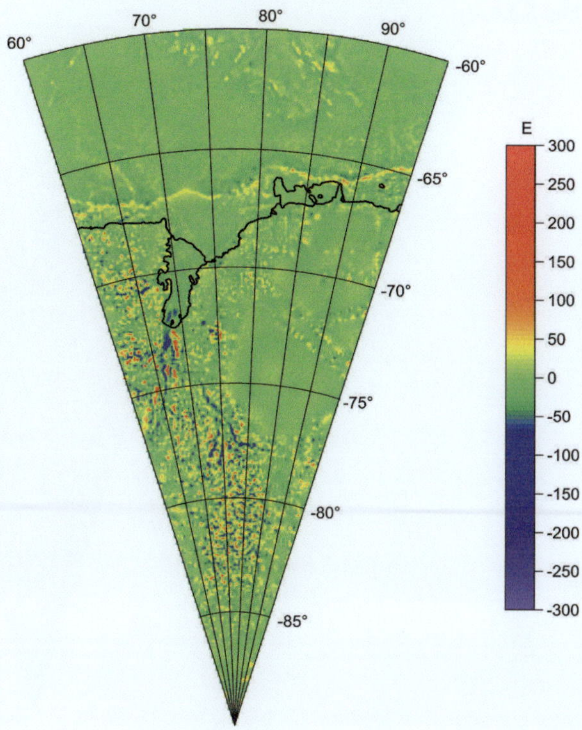

Fig. 5.3.3 I_1

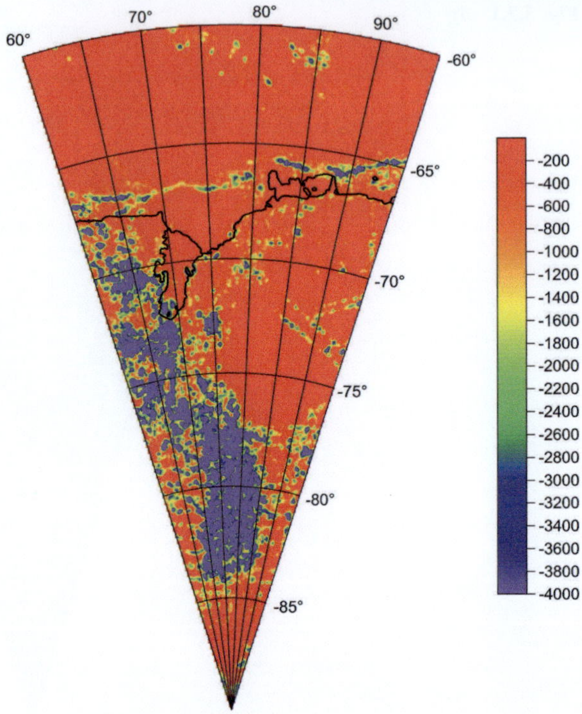

Fig. 5.3.4 I_2

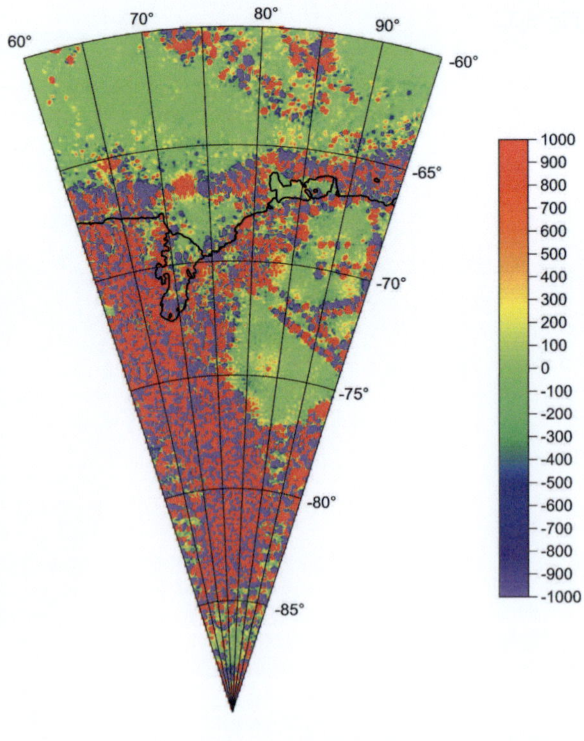

Fig. 5.3.5 I

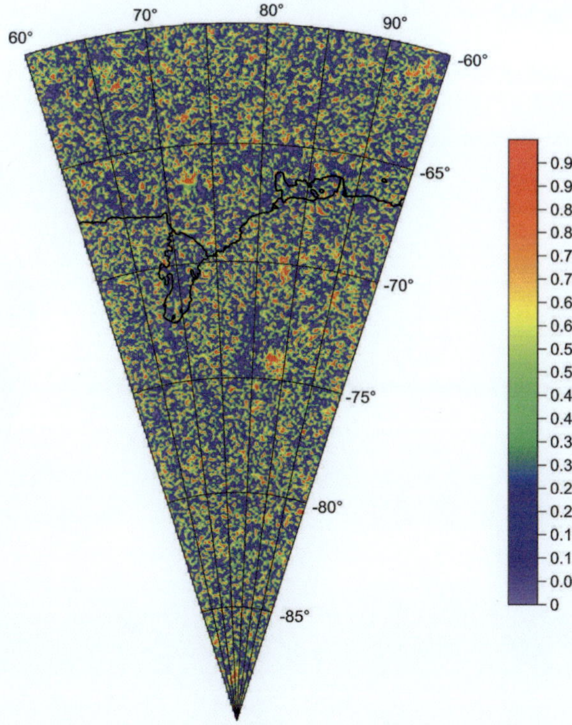

Fig. 5.3.6 θ

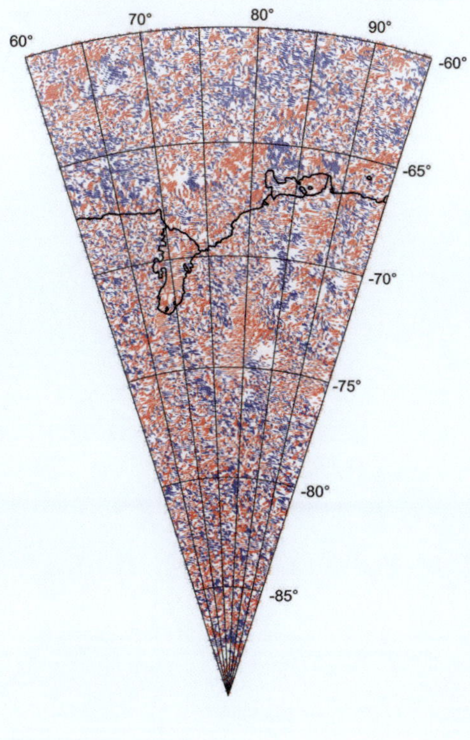

Fig. 5.3.7 *vd*

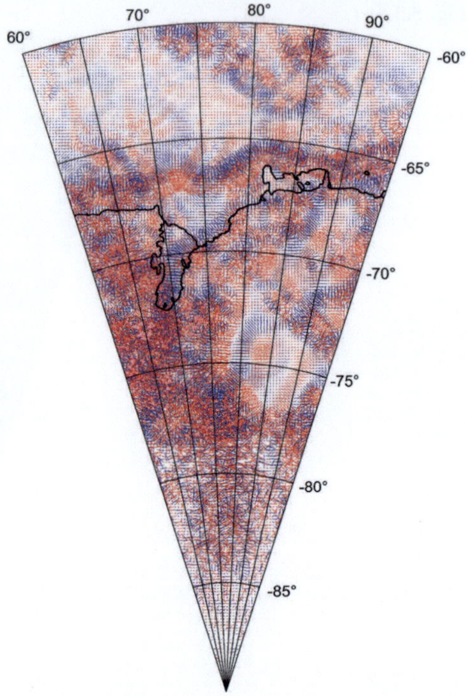

Fig. 5.3.8 *vd,* dilatation

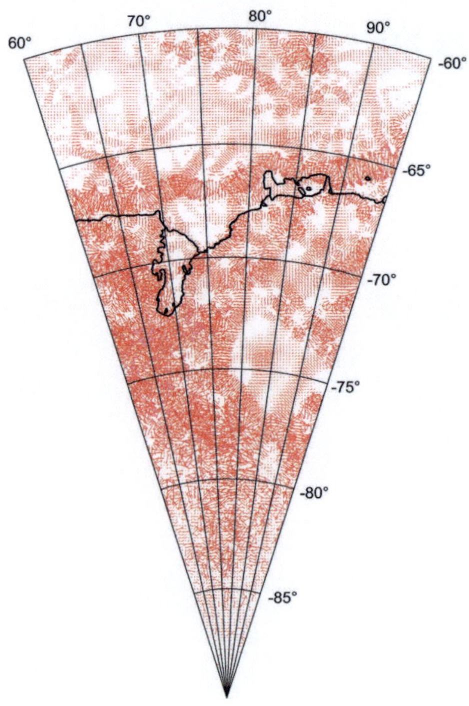

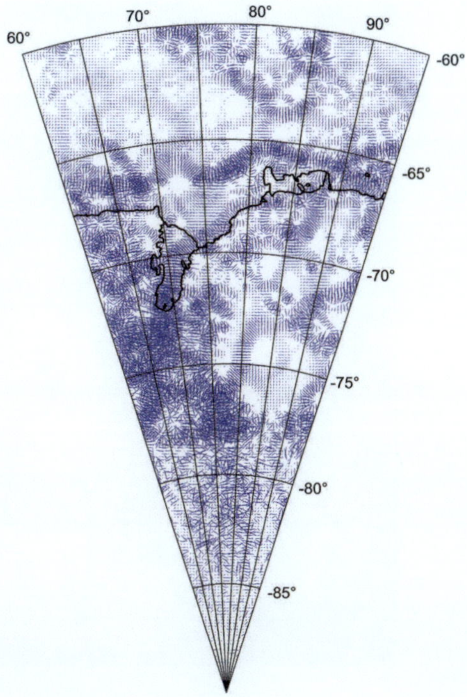

Fig. 5.3.9 *vd,* compression

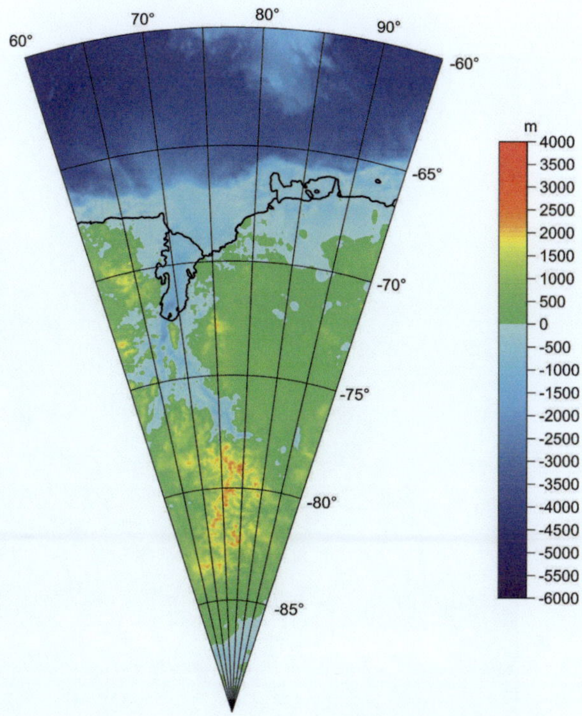

Fig. 5.3.10 Bedmap 2 topography

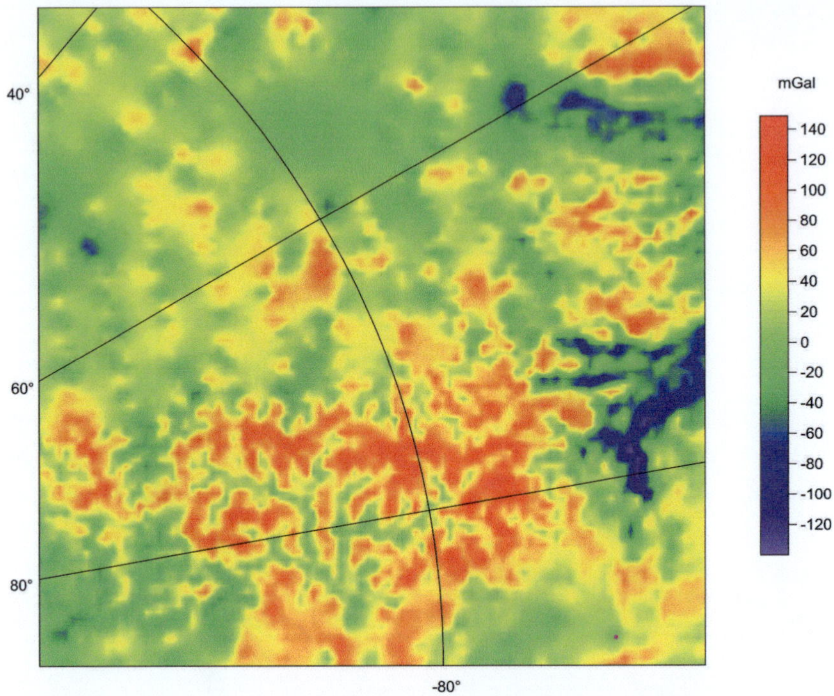

Fig. 5.3.11 *Δg*. Gamburtsev Mountain Range (completely under the ice cover)

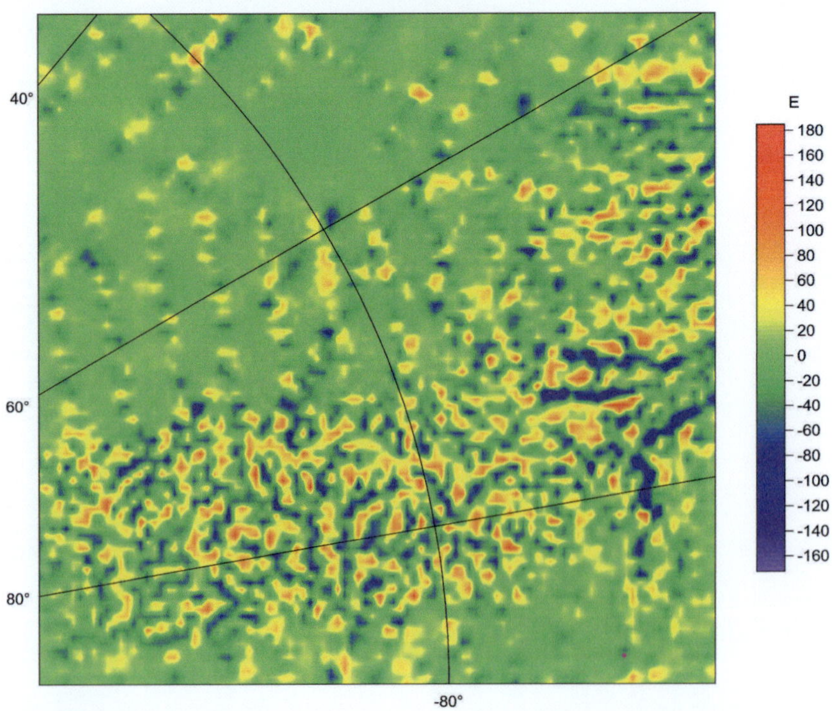

Fig. 5.3.12 T_{zz}

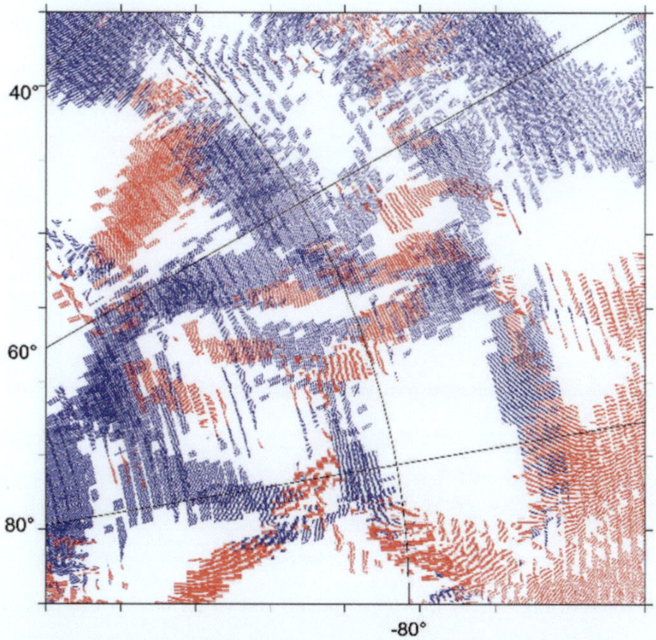

Fig. 5.3.13 Strike angles for $I < 0.3$

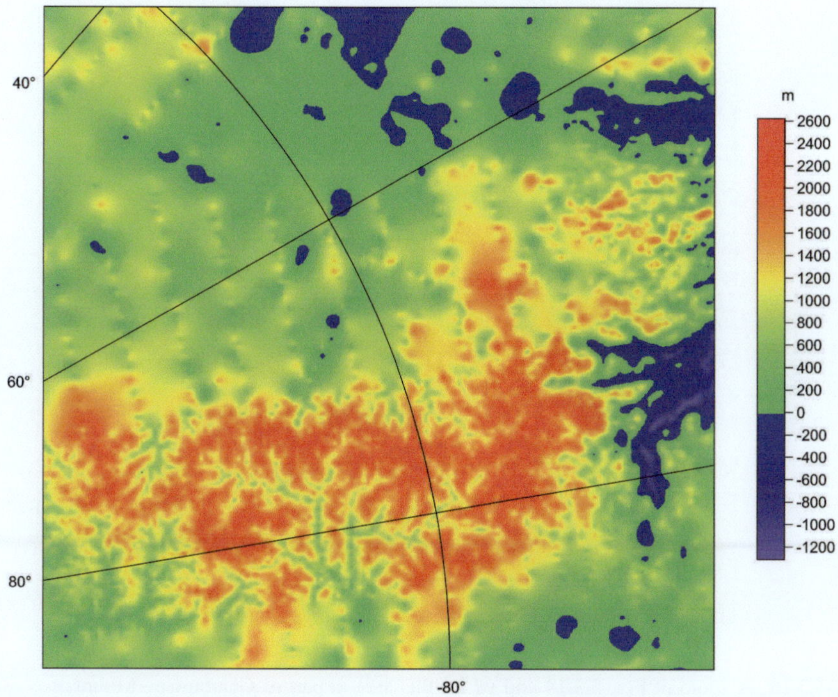

Fig. 5.3.14 Bedrock topography (Bedmap 2) of Gamburtsev Mountains

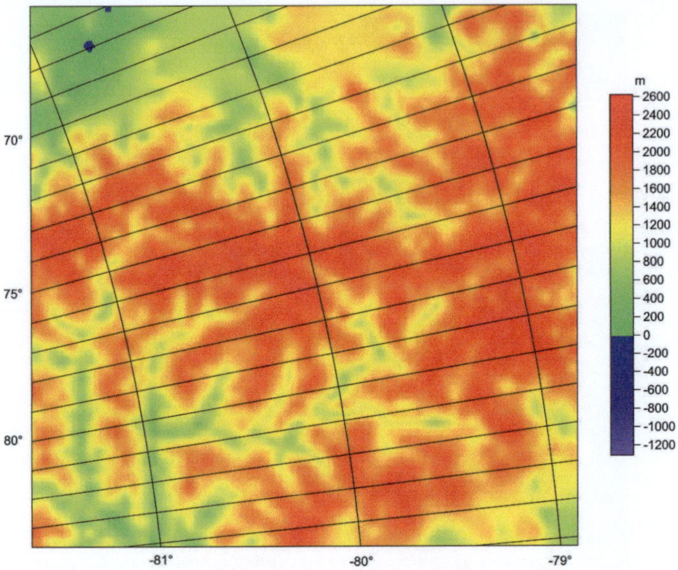

Fig. 5.3.15 A zoom of Fig. 5.3.14 (*the upper panel*) and *vd* for this area (a part of Gamburtsev Mountains)

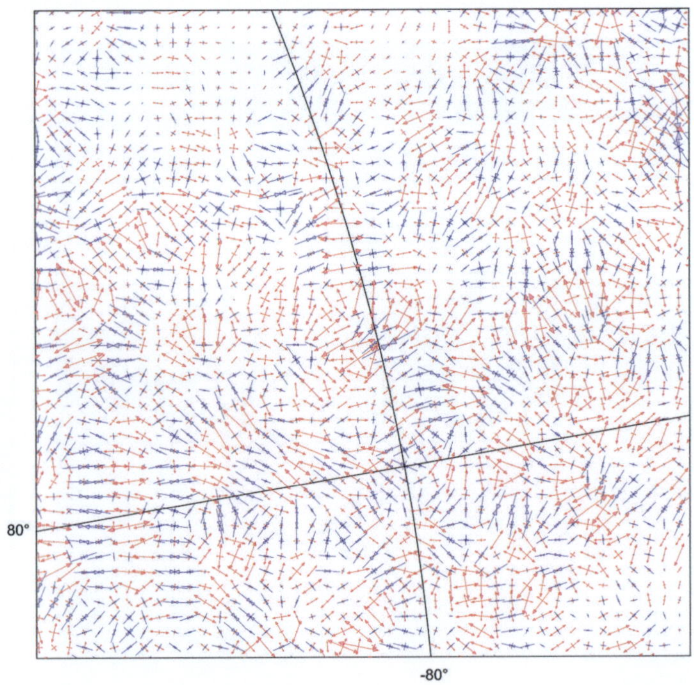

Fig. 5.3.16 A zoom of Fig. 5.3.14 and *vd* for this area (a part of Gamburtsev Mountains)

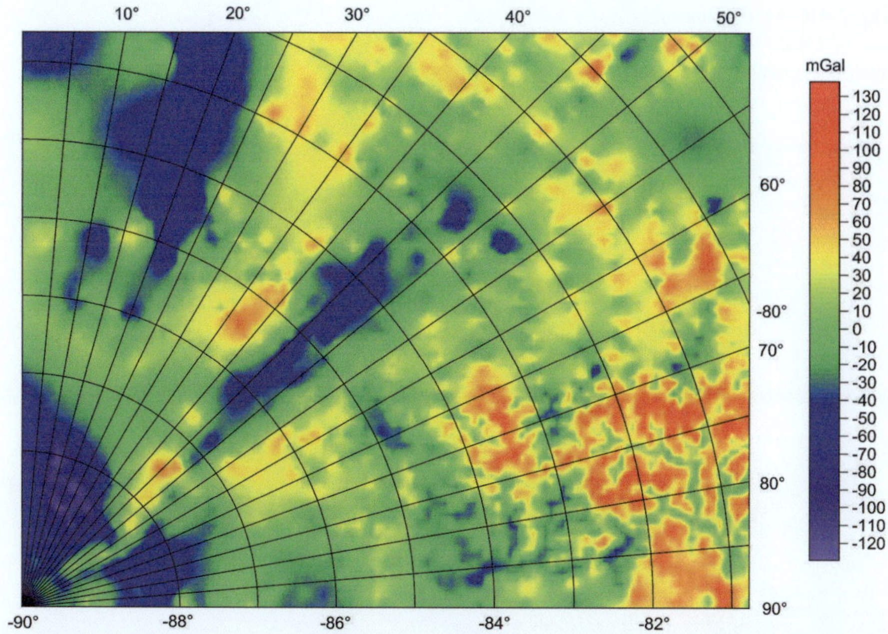

Fig. 5.3.17 A close-up to Δg (*the upper panel*) for the southern part of Gamburtsev Mountains

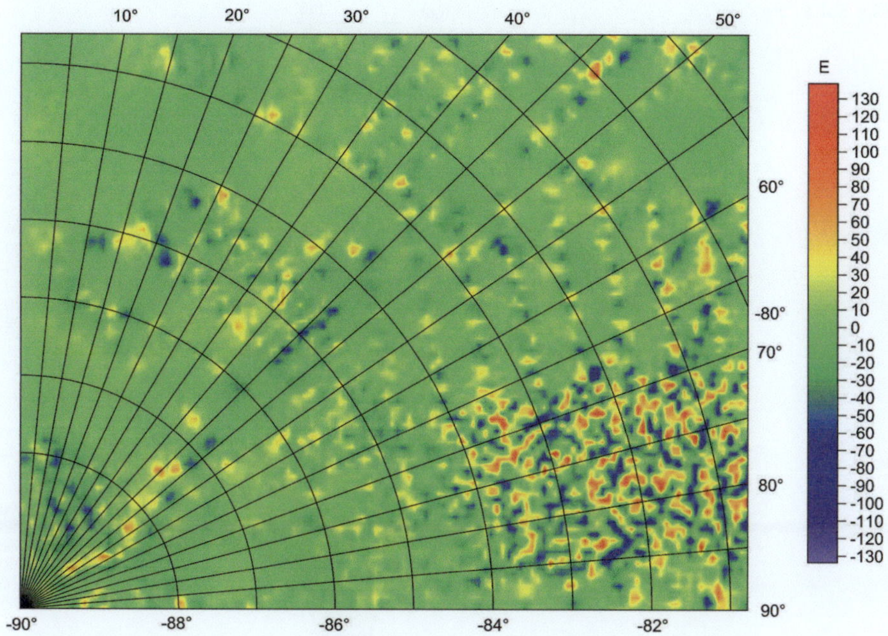

Fig. 5.3.18 A close-up to T_{zz} for the southern part of Gamburtsev Mountains

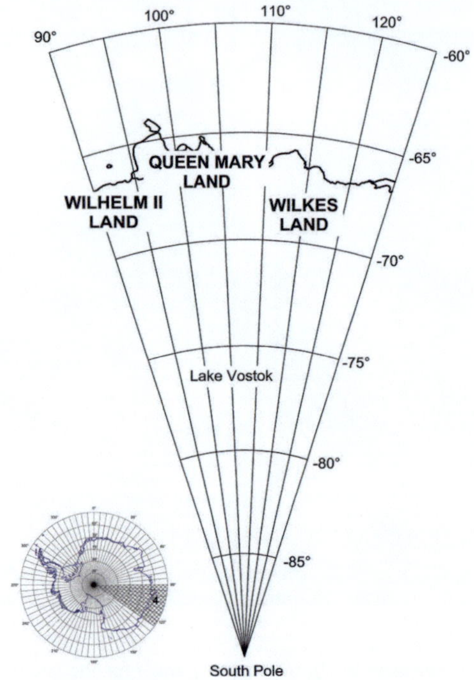

Fig. 5.4.0 An orientation rose and introductory figure for this segment

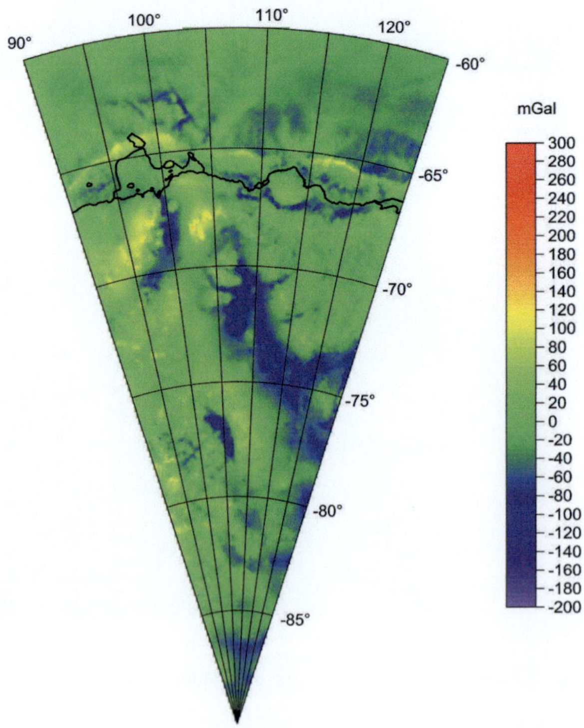

Fig. 5.4.1 Δg

Fig. 5.4.2 T_{zz}

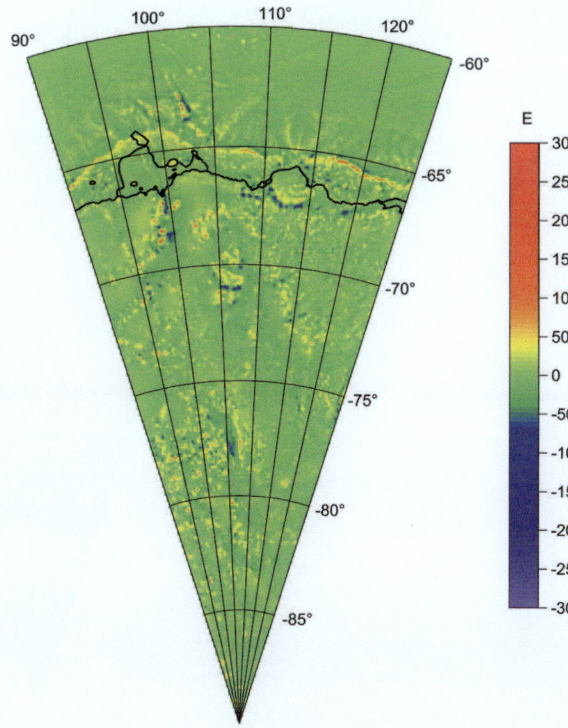

Fig. 5.4.3 I_1

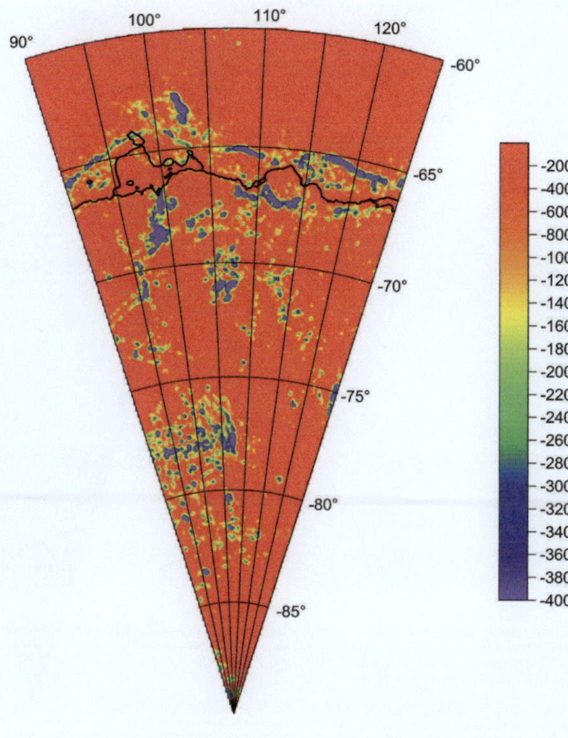

Fig. 5.4.4 I_2

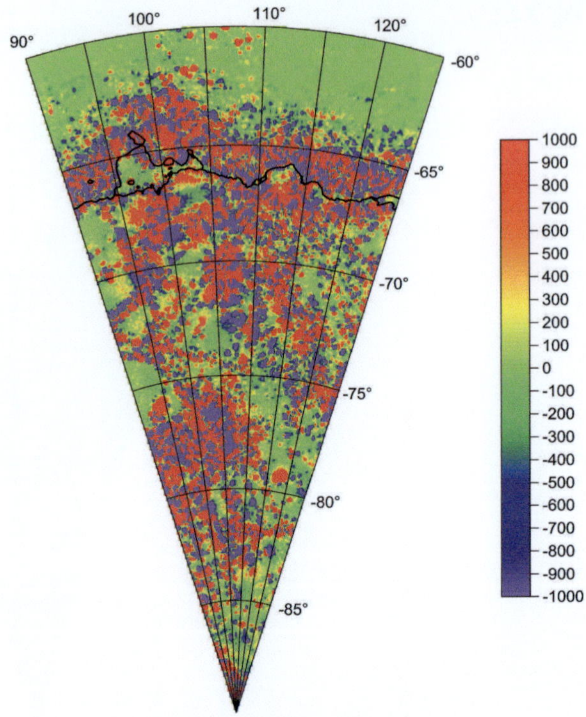

Fig. 5.4.5 I

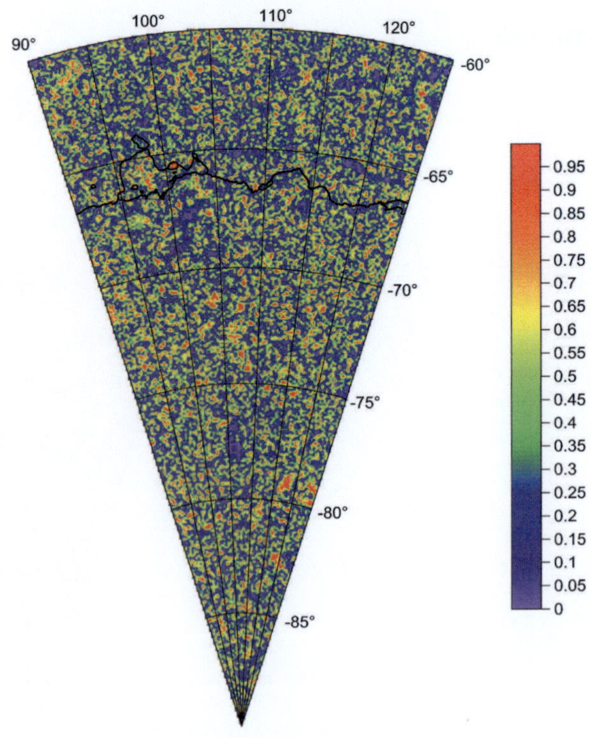

Fig. 5.4.6 θ

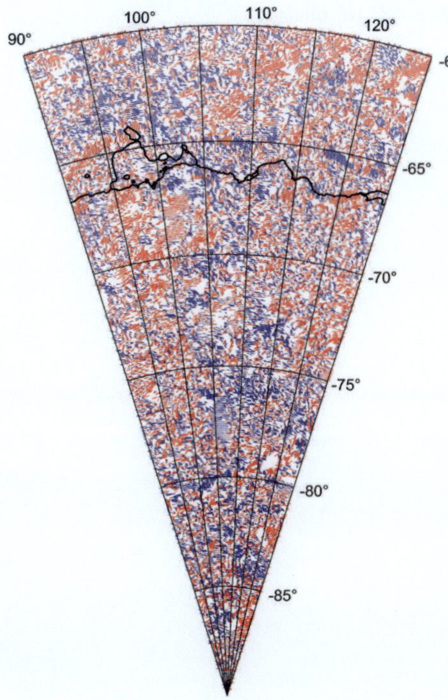

Fig. 5.4.7 *vd*

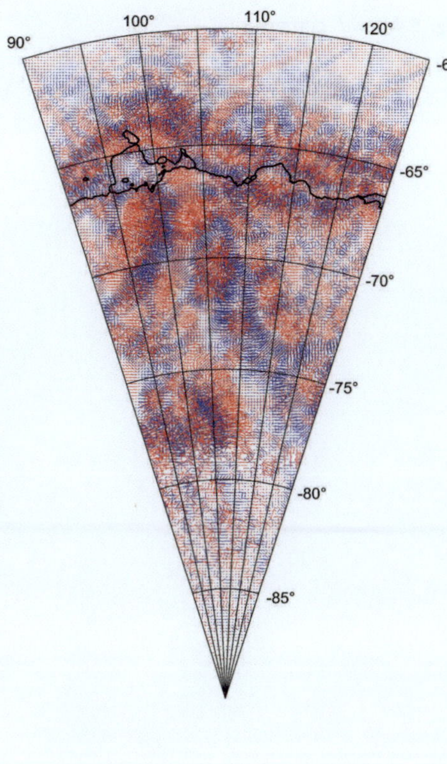

Fig. 5.4.8 *vd,* dilatation

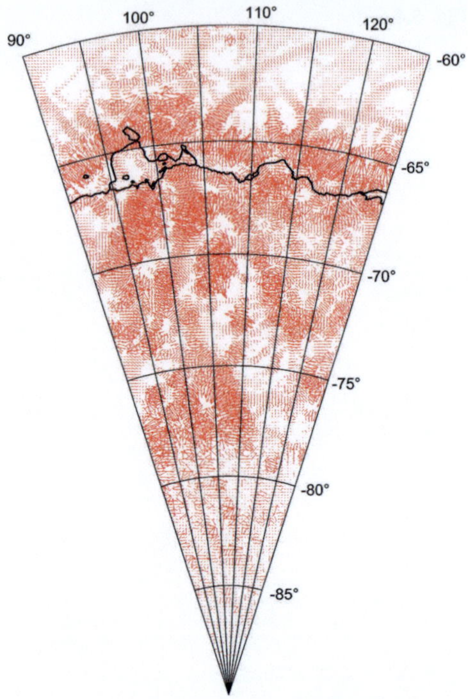

Fig. 5.4.9 *vd,* compression

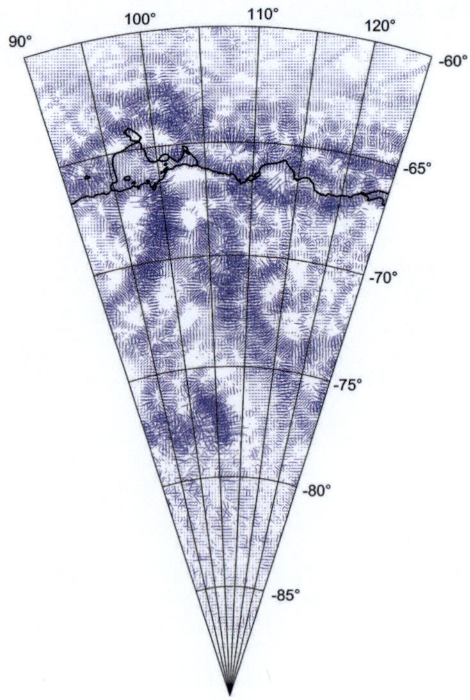

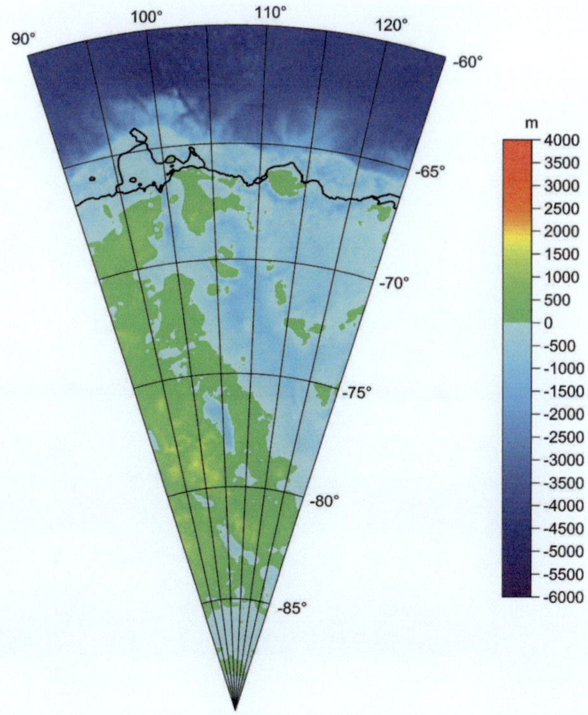

Fig. 5.4.10 Bedmap 2 topography

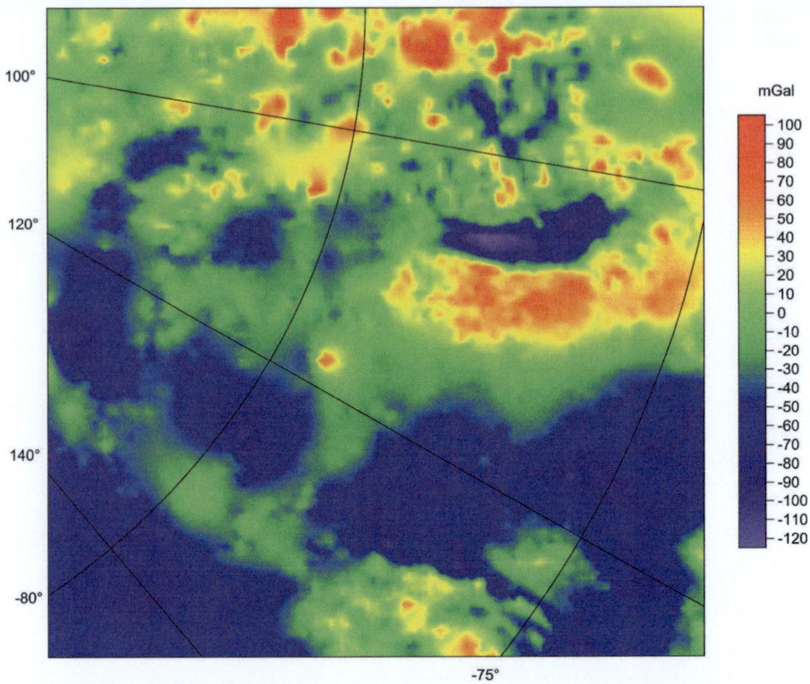

Fig. 5.4.11 Δg. A zoom for the area of Lake Vostok (elongated strong negative anomaly on right up) and a part of Wilkes low-land

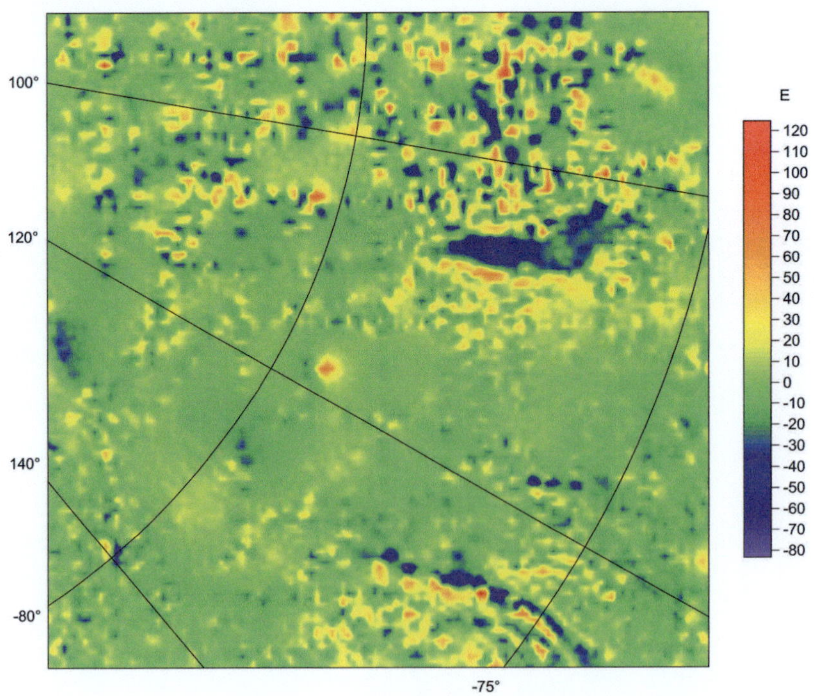

Fig. 5.4.12 T_{zz}. A zoom for the area of Lake Vostok

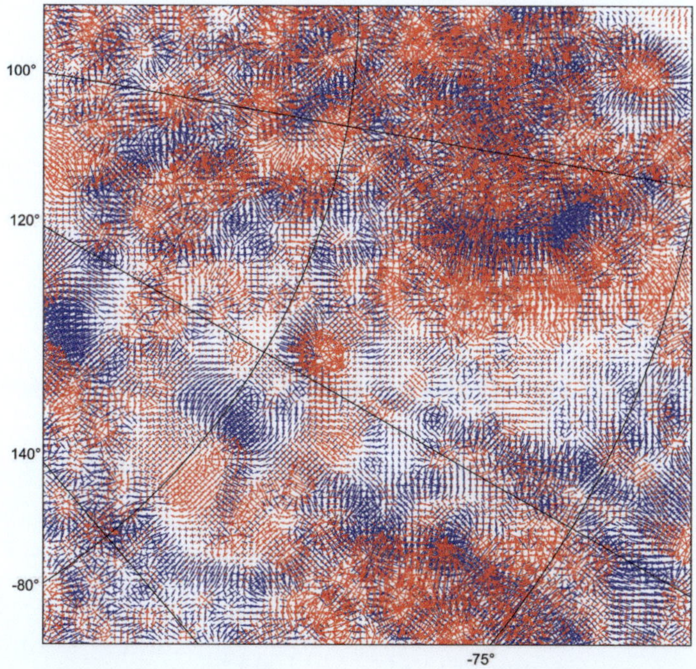

Fig. 5.4.13 The *vd* in the Lake Vostok area. Note the local extrema values of Δg, T_z and *vd* in Figs. 5.4.11–5.4.13 at $\varphi = \sim -79°$, $\lambda = \sim 119°$E, surrounded by relatively flat vicinity (for more information see Klokočník et al. 2016)

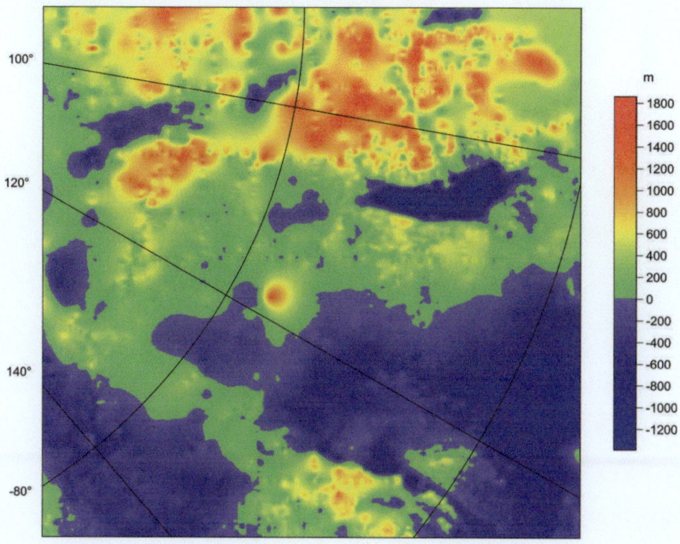

Fig. 5.4.14 Bedrock topography (with Bedmap 2) for the Lake Vostok area. For more information about this area, see Klokočník et al. (2016)

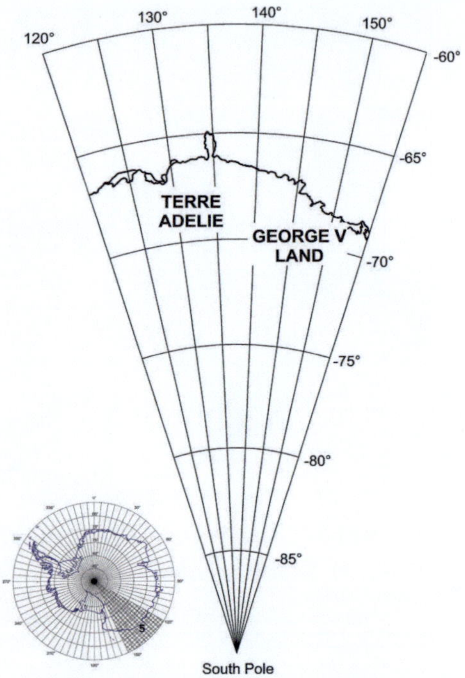

Fig. 5.5.0 An orientation rose and introductory figure for this segment showing a few topographic features of the area

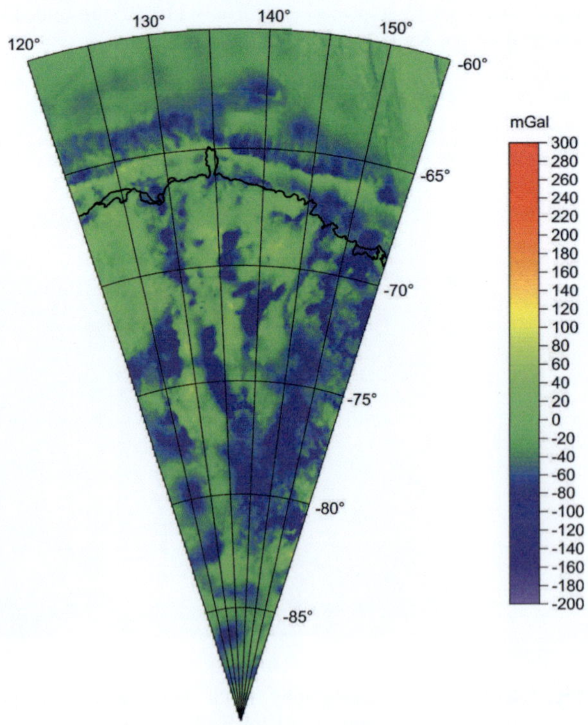

Fig. 5.5.1 Δg

Fig. 5.5.2 T_{zz}

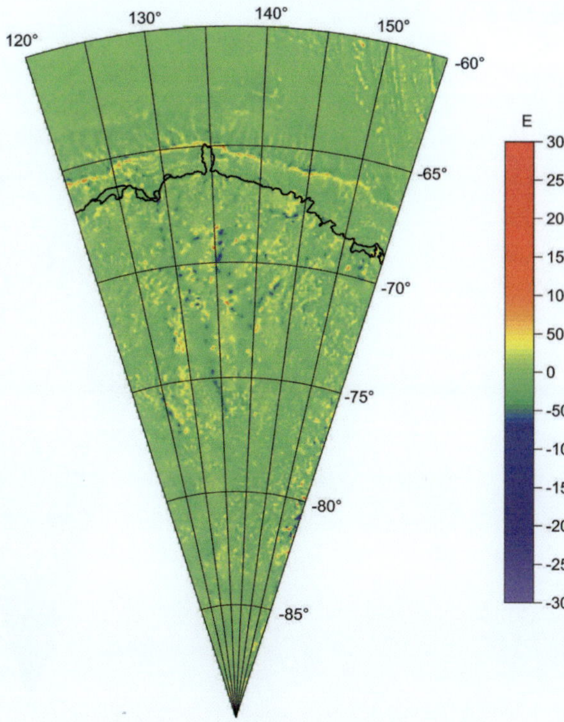

Fig. 5.5.3 I_1

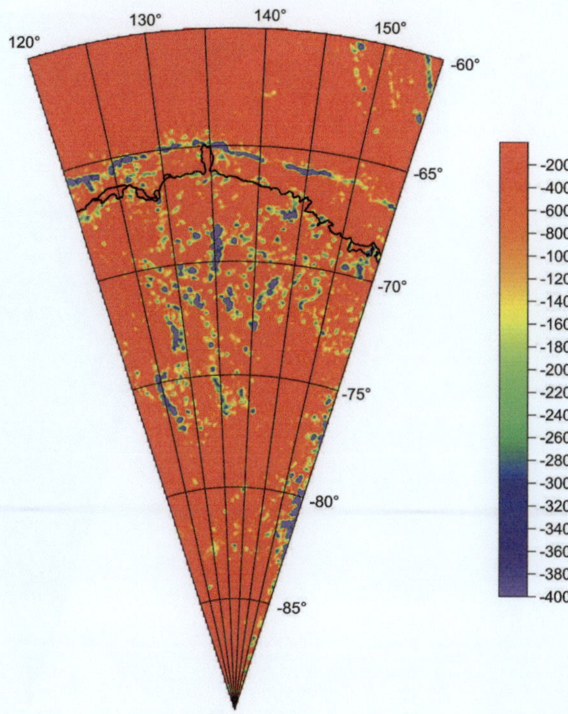

Fig. 5.5.4 I_2

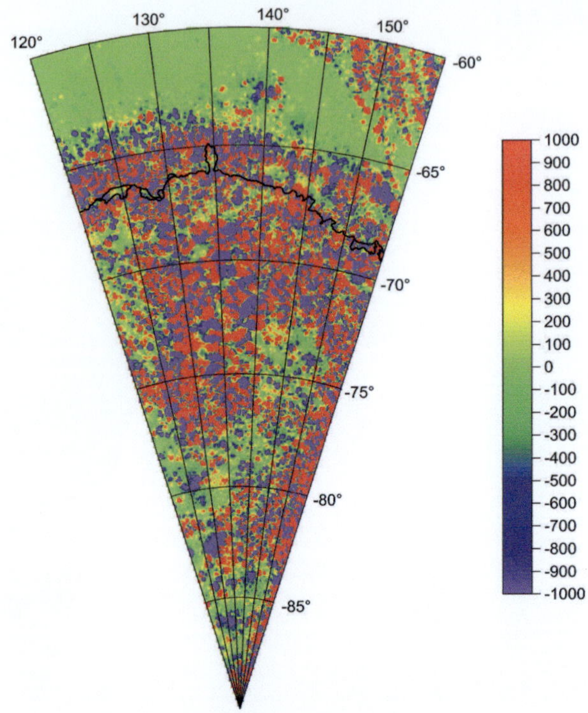

Fig. 5.5.5 I

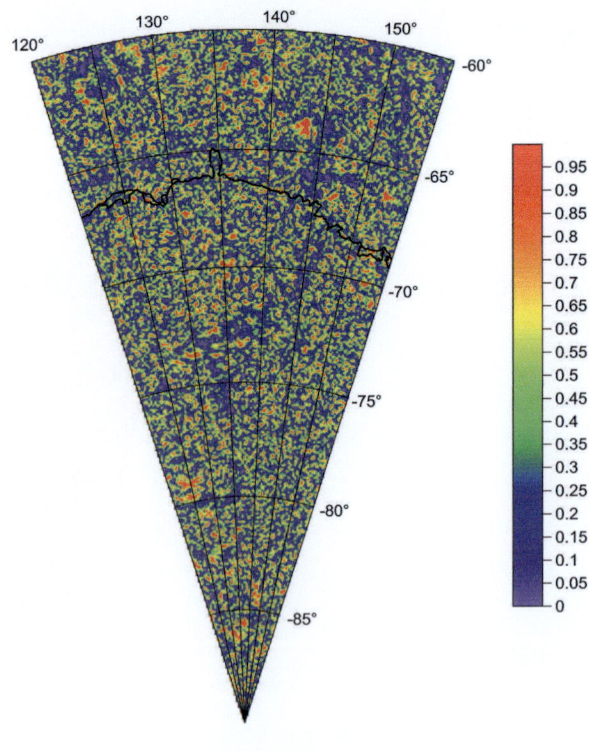

Fig. 5.5.6 θ

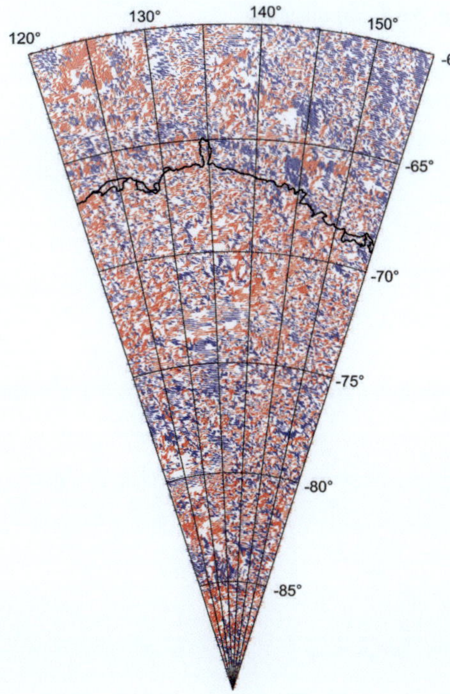

Fig. 5.5.7 *vd*

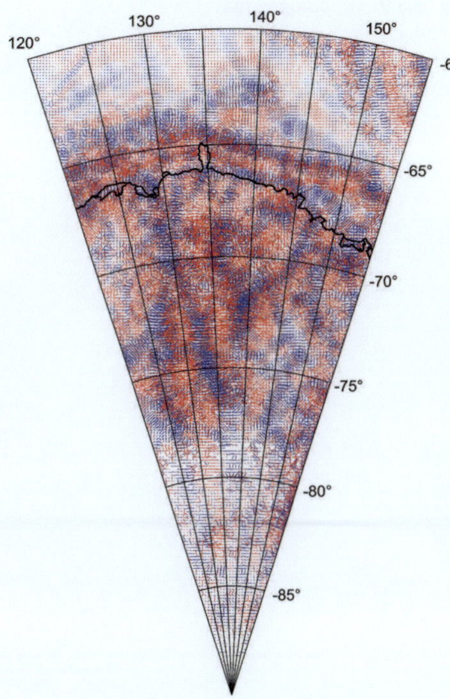

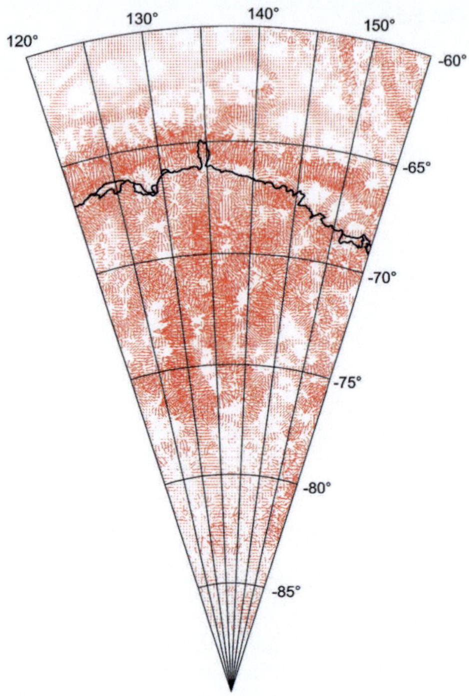

Fig. 5.5.8 *vd,* dilatation

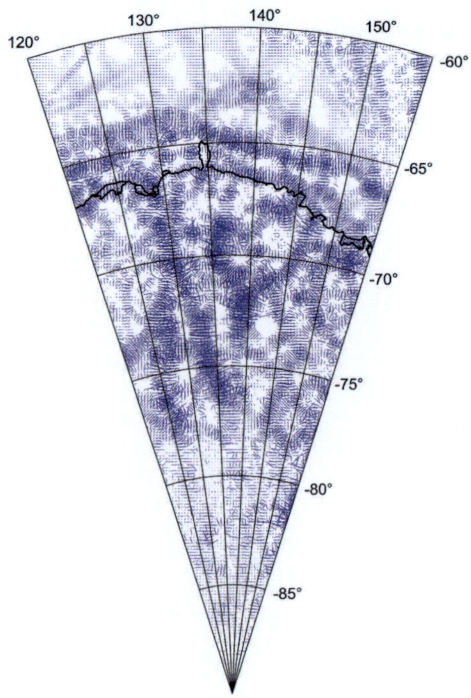

Fig. 5.5.9 *vd,* compression

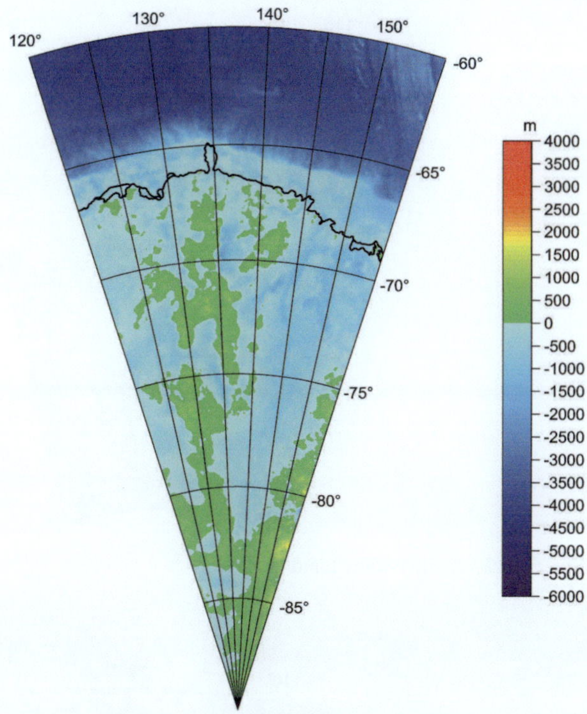

Fig. 5.5.10 Bedmap 2 topography

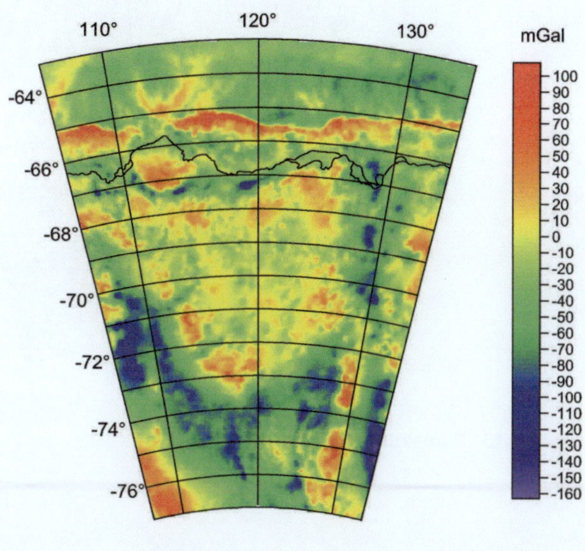

Fig. 5.5.11 Δg in Wilkes Land

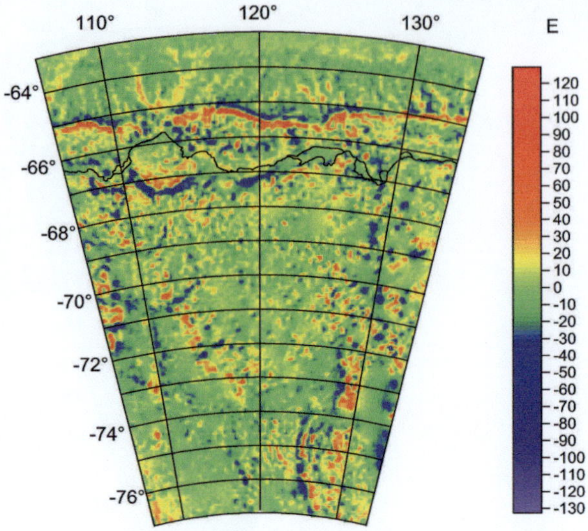

Fig. 5.5.12 T_{zz} in Wilkes Land

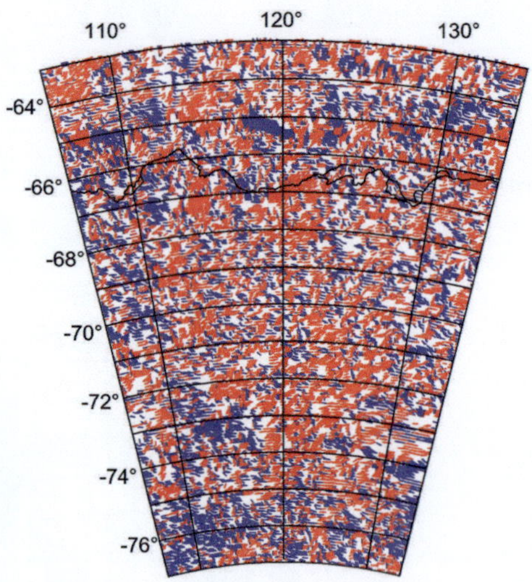

Fig. 5.5.13 θ in Wilkes Land

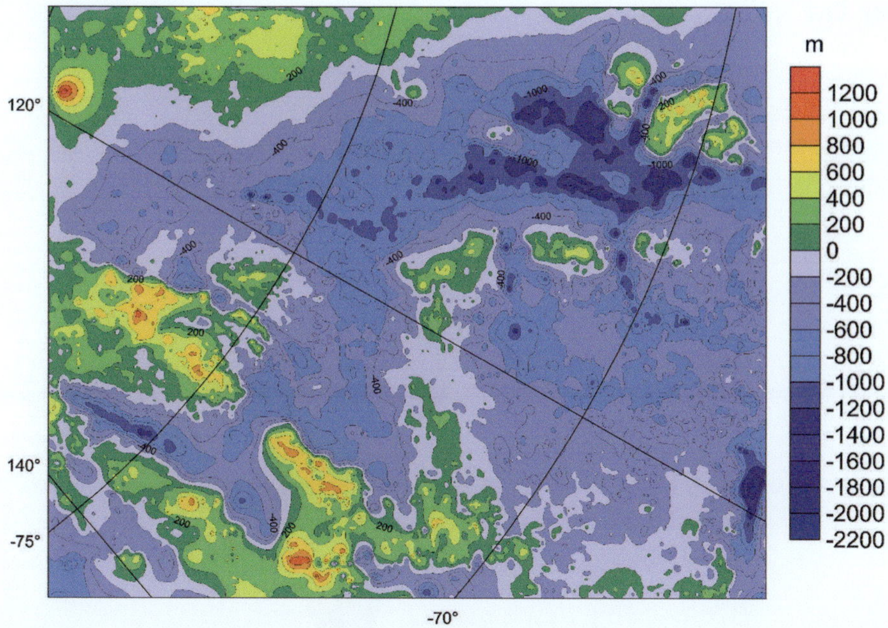

Fig. 5.5.14 Bedrock topography (Bedmap 2) for Wilkes Land

Fig. 5.6.0 An orientation rose and introductory figure for this segment showing a few topographic features of the area

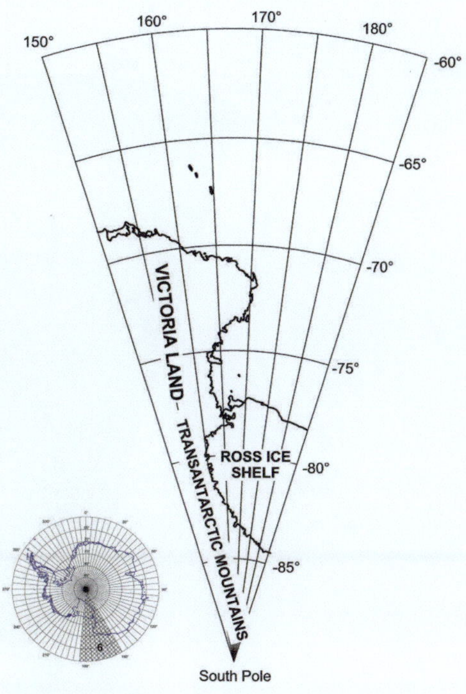

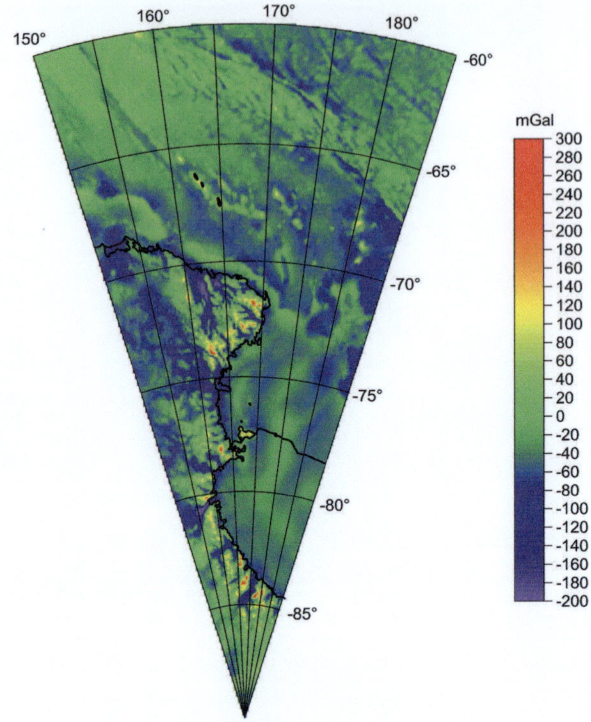

Fig. 5.6.1 Δg

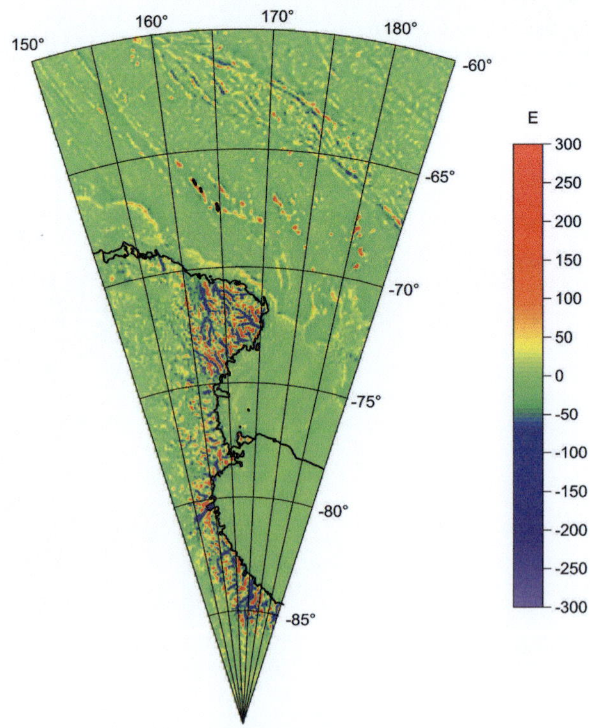

Fig. 5.6.2 T_{zz}

Fig. 5.6.3 I_1

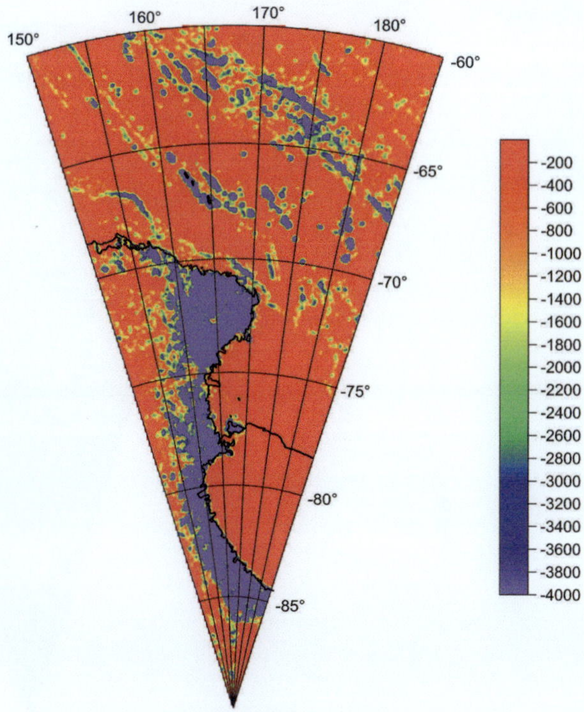

Fig. 5.6.4 I_2

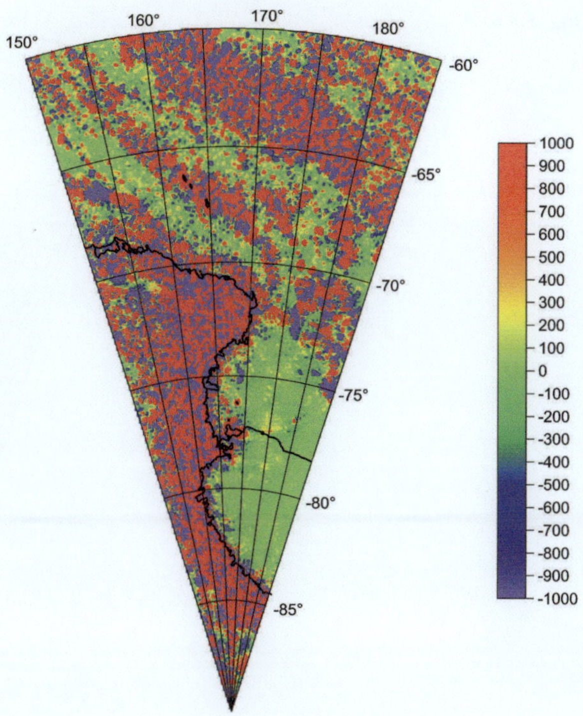

Fig. 5.6.5 I

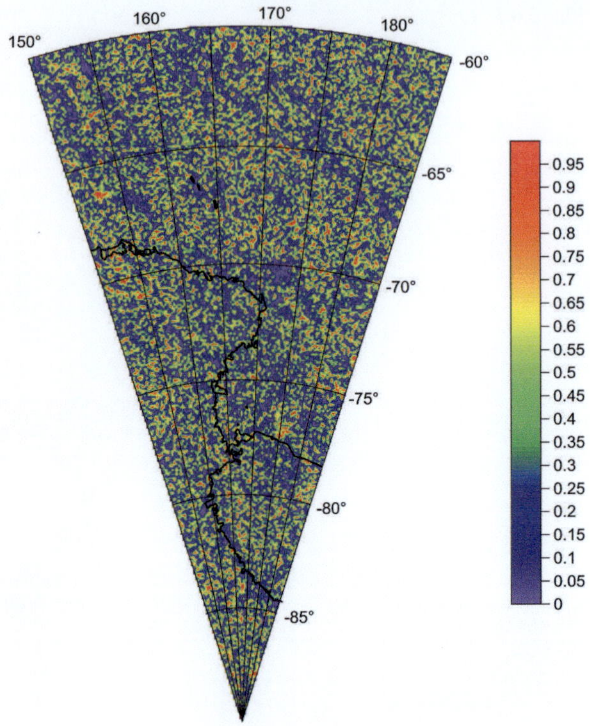

Fig. 5.6.6 θ

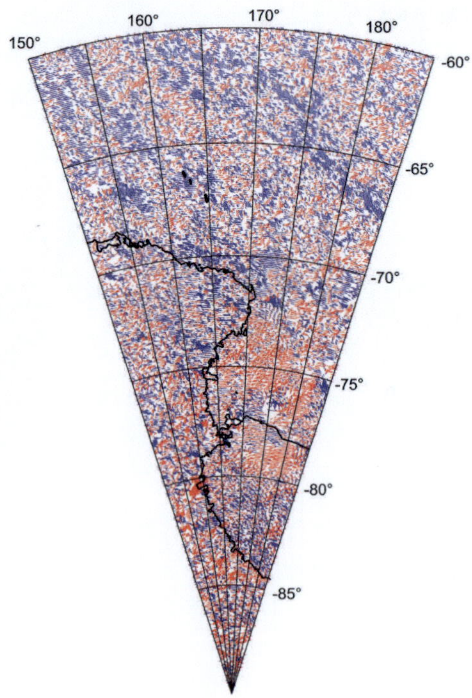

Fig. 5.6.7 *vd*

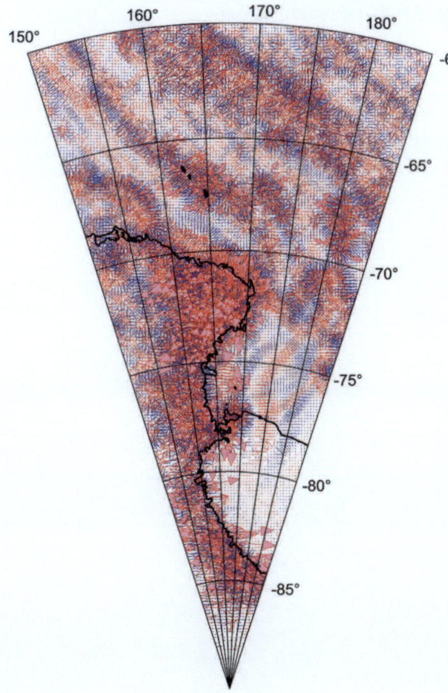

Fig. 5.6.8 *vd,* dilatation

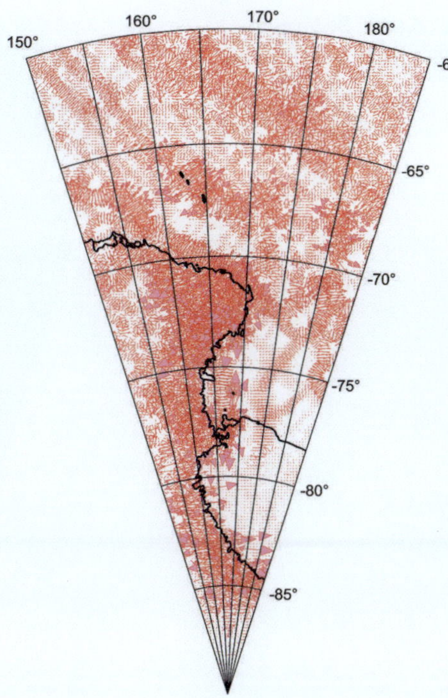

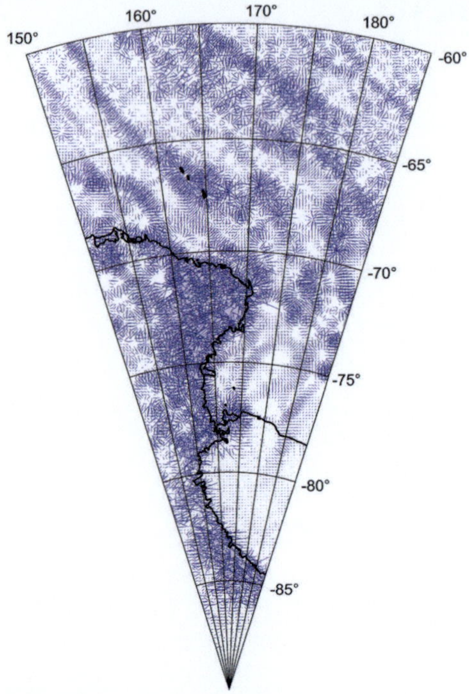

Fig. 5.6.9 *vd,* compression

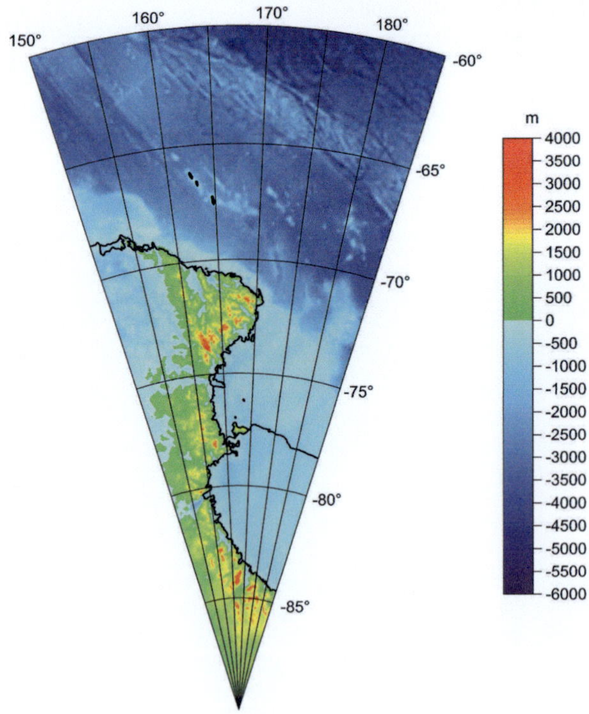

Fig. 5.6.10 Bedmap 2 topography

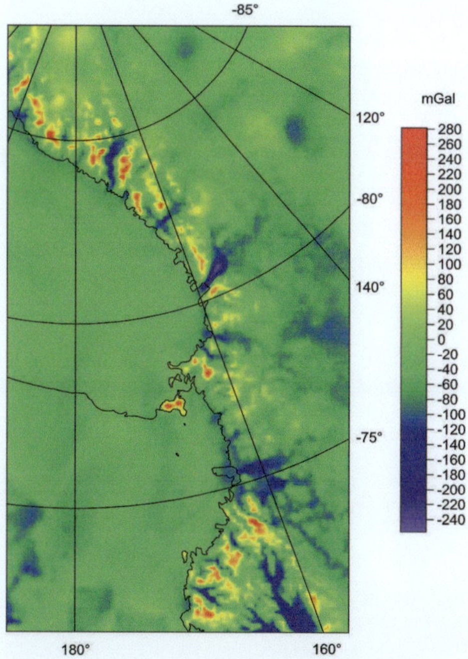

Fig. 5.6.11 Δg. A part of the Transantarctic Mountains, the Ross Ice Shelf and Ross Island

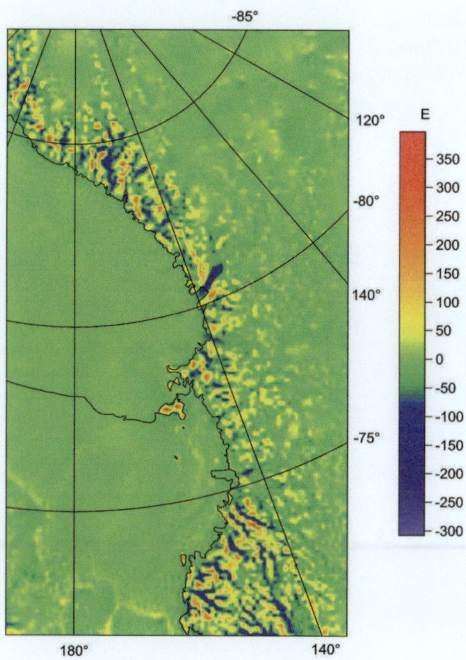

Fig. 5.6.12 T_{zz}

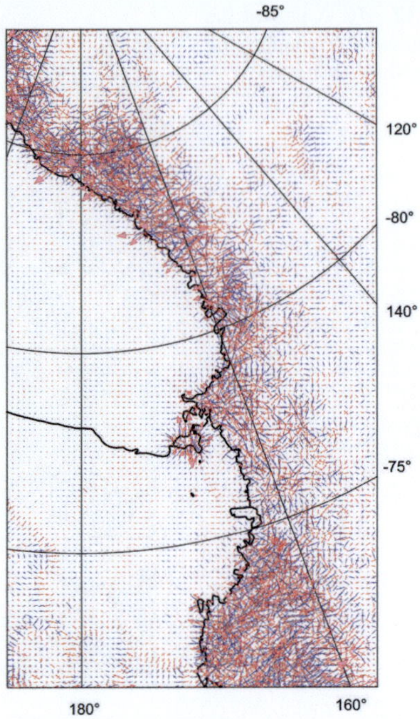

Fig. 5.6.13 *vd*

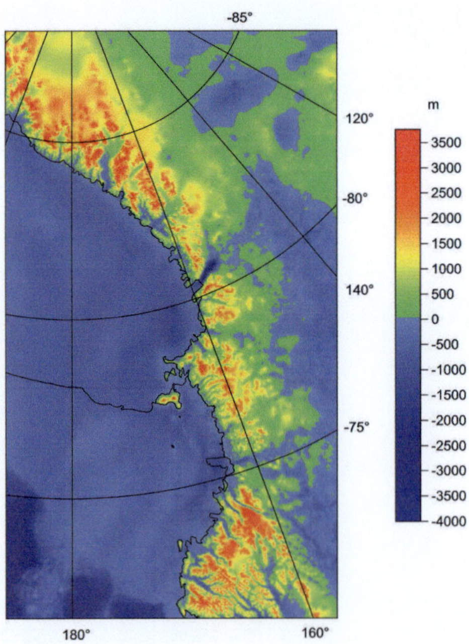

Fig. 5.6.14 Bedrock topography (Bedmap 2) of a part of the Transantarctic Mountains

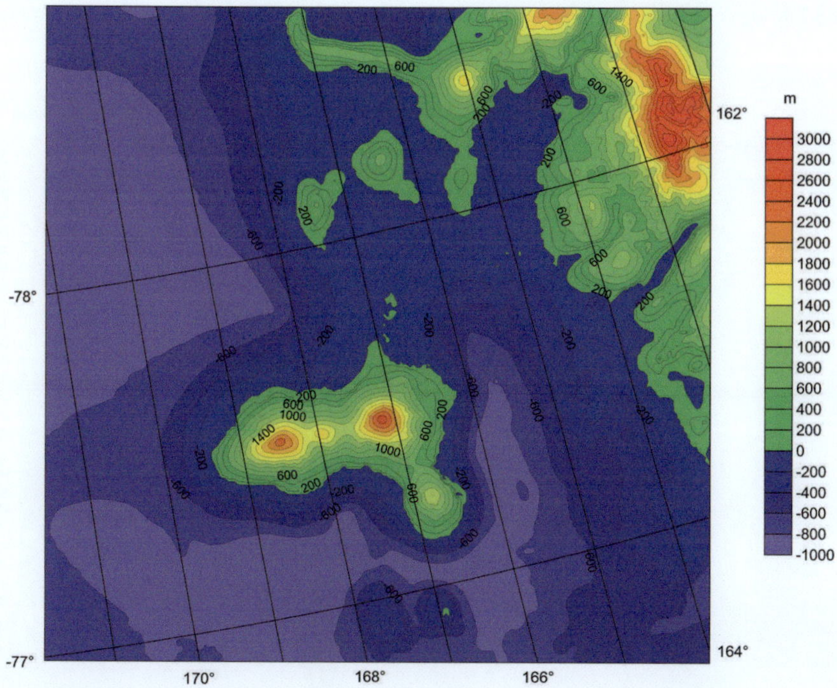

Fig. 5.6.15 Bedmap 2 topography for the Ross Island

Fig. 5.7.0 An orientation rose and introductory figure for this segment showing a few topographic features of the area

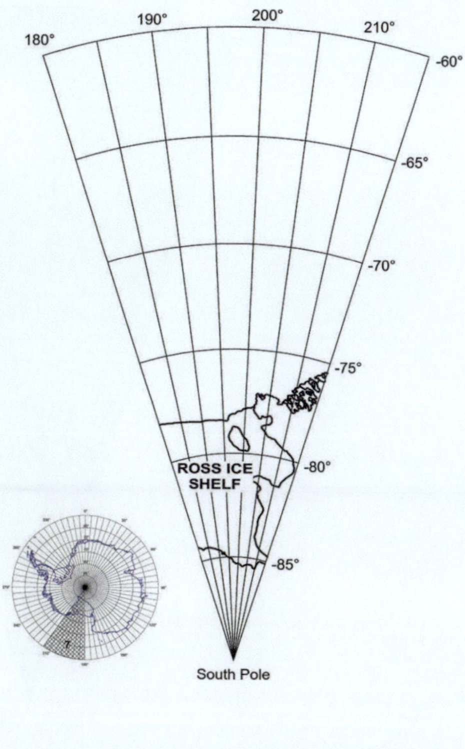

Fig. 5.7.1 Δg

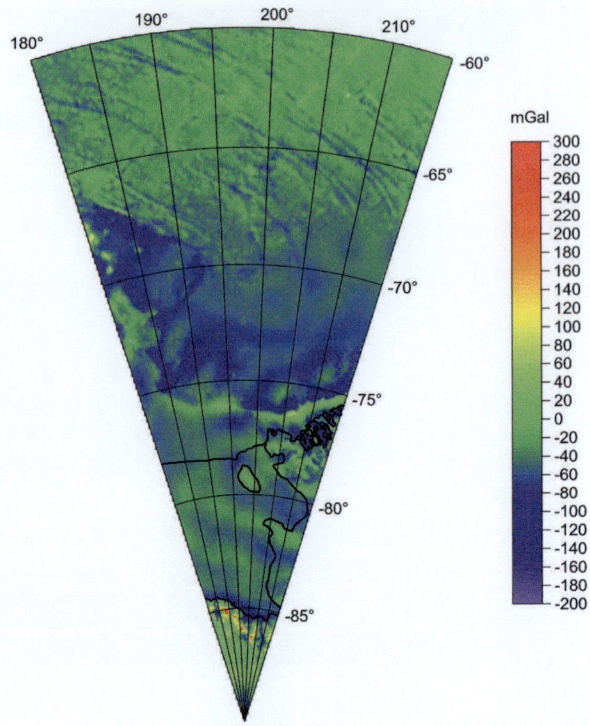

Fig. 5.7.2 T_{zz}

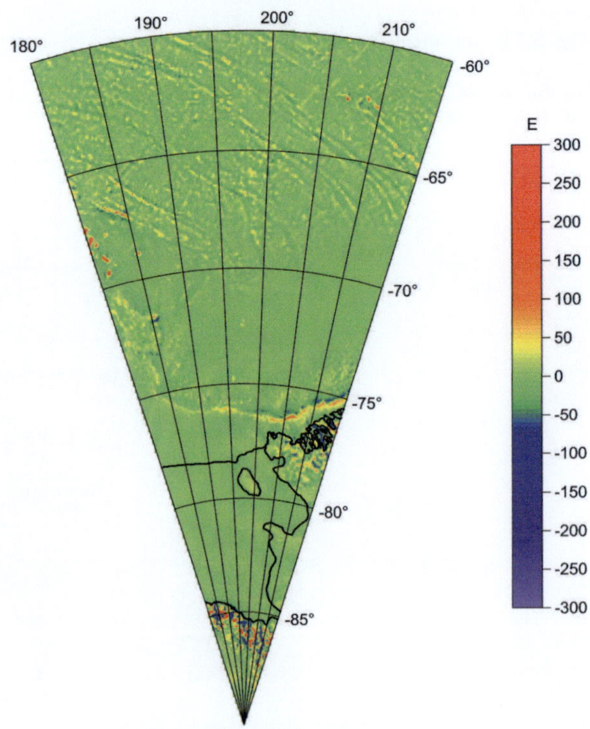

Fig. 5.7.3 I_1

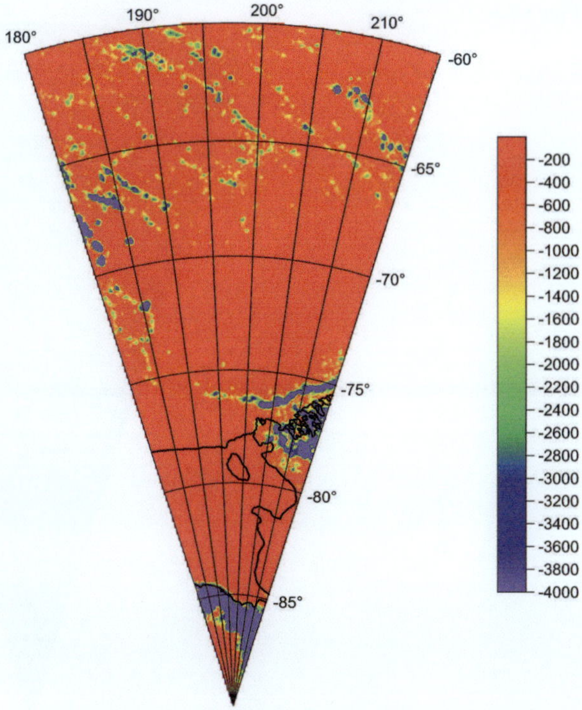

Fig. 5.7.4 I_2

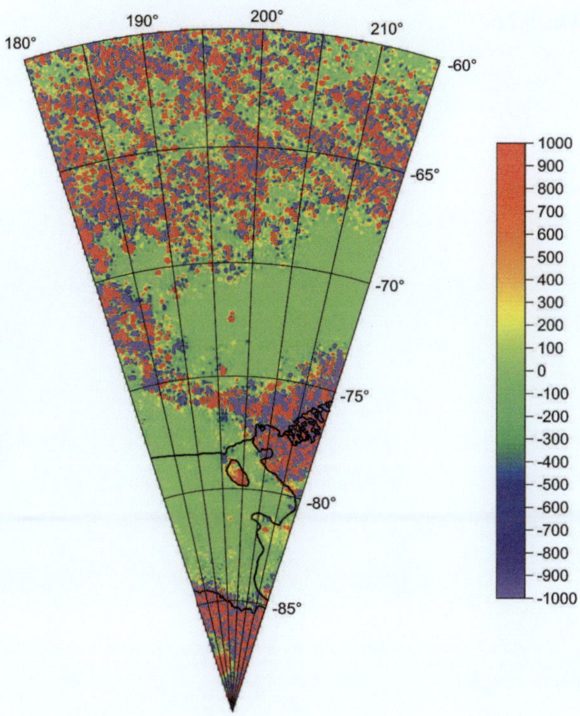

Fig. 5.7.5 *I*

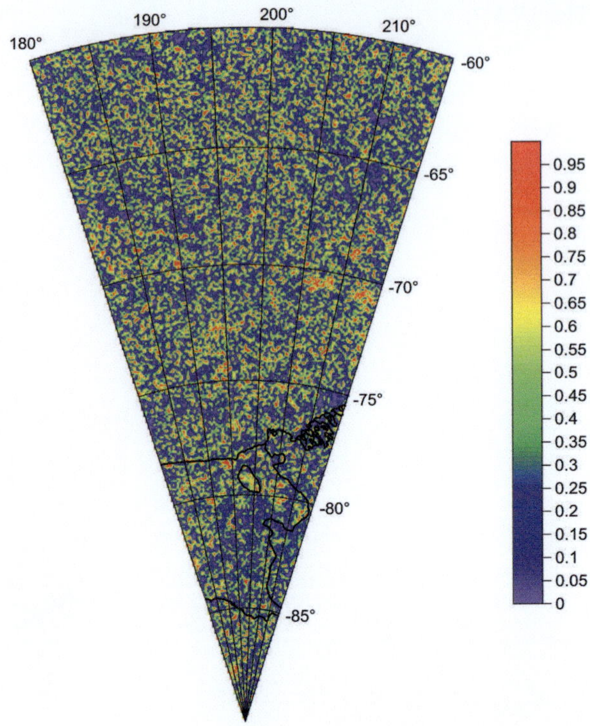

Fig. 5.7.6 *θ*

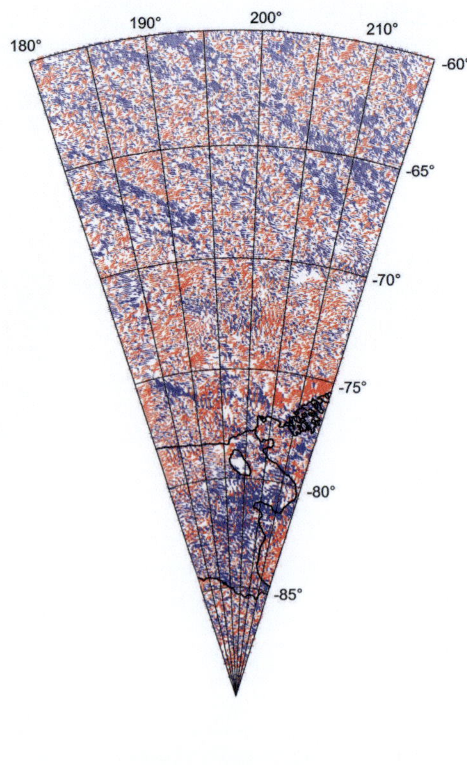

Fig. 5.7.7 *vd*

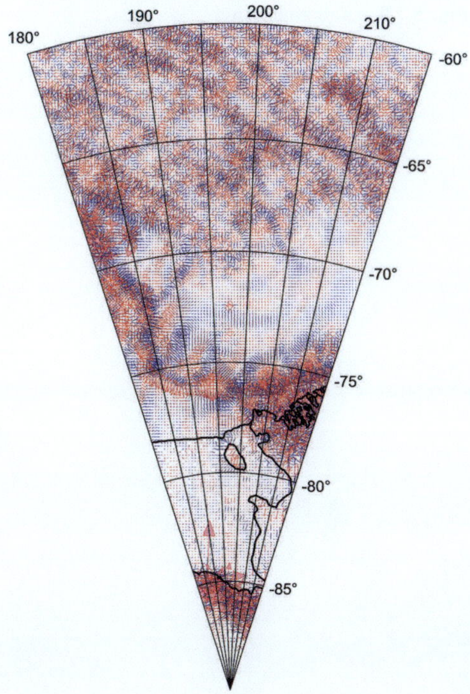

Fig. 5.7.8 *vd,* dilatation

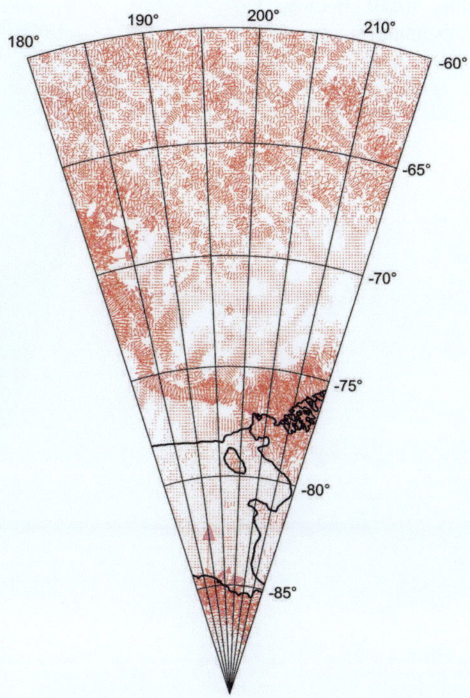

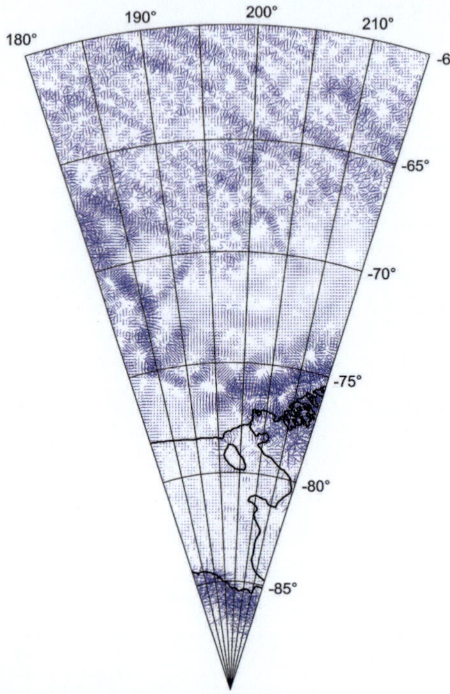

Fig. 5.7.9 *vd*, compression

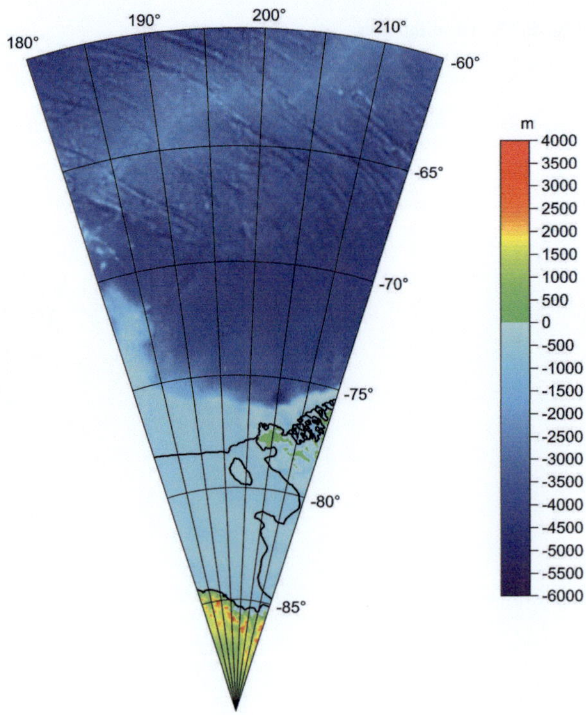

Fig. 5.7.10 Bedmap 2 topography

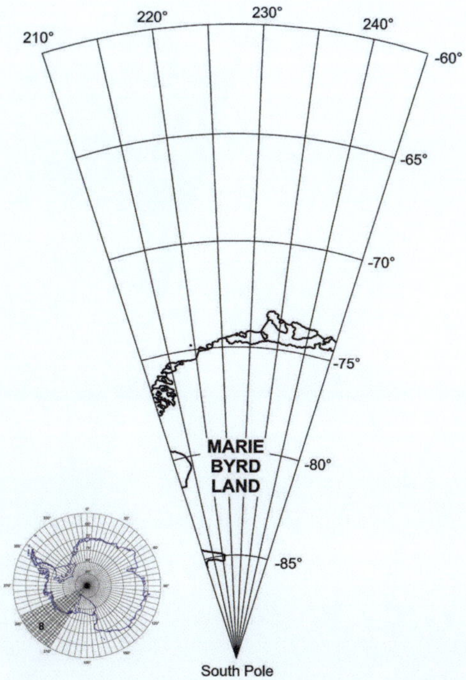

Fig. 5.8.0 An orientation rose and introductory figure for this segment

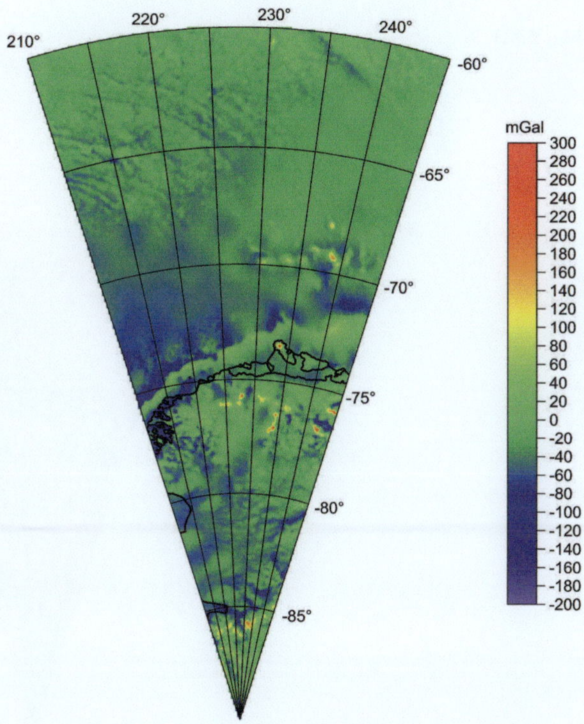

Fig. 5.8.1 Δg

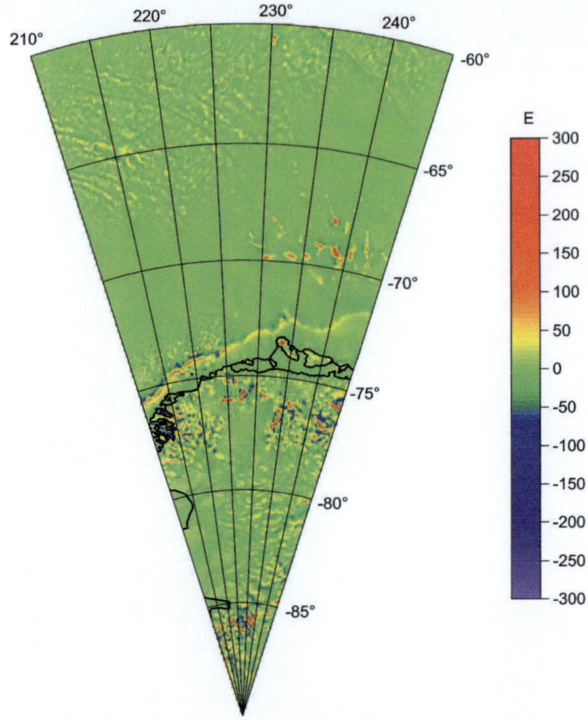

Fig. 5.8.2 T_{zz}

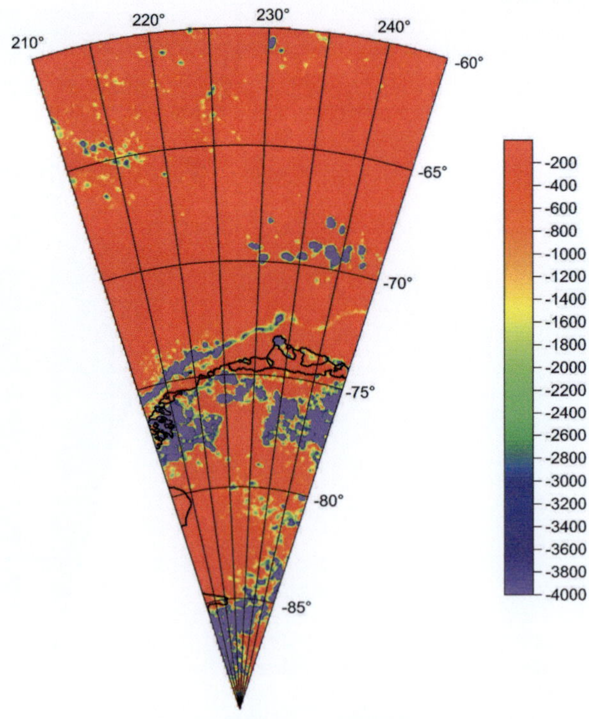

Fig. 5.8.3 I_1

Fig. 5.8.4 I_2

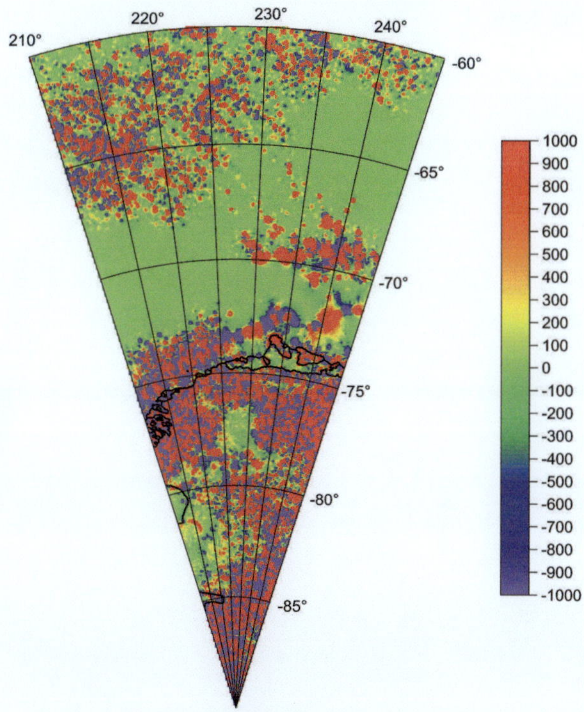

Fig. 5.8.5 I

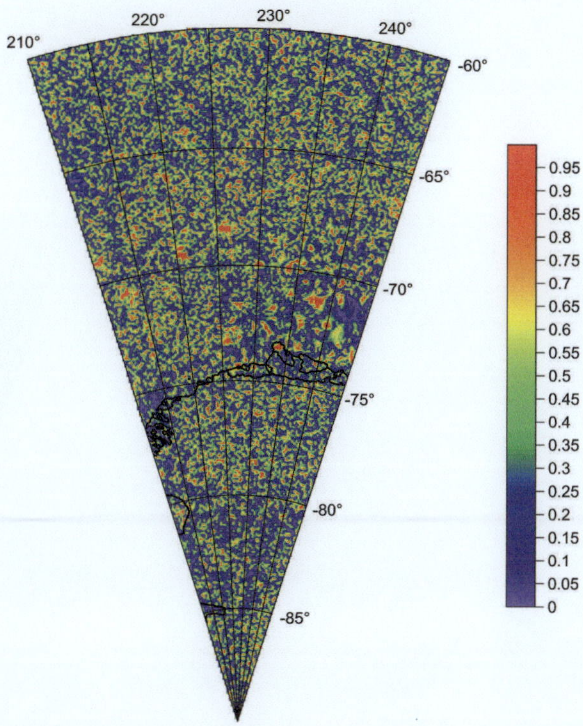

Fig. 5.8.6 θ

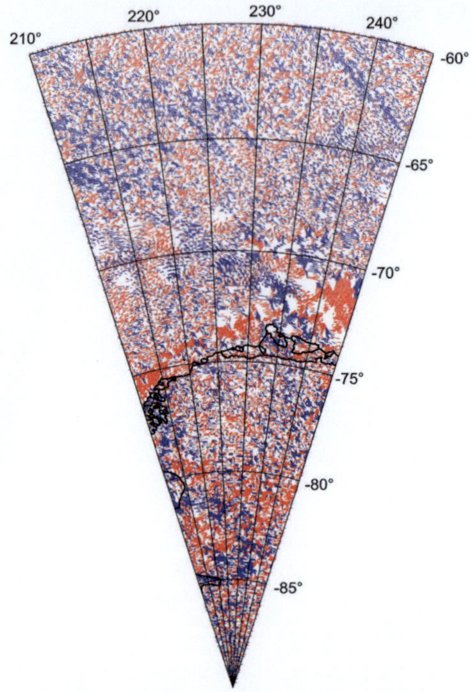

Fig. 5.8.7 *vd*

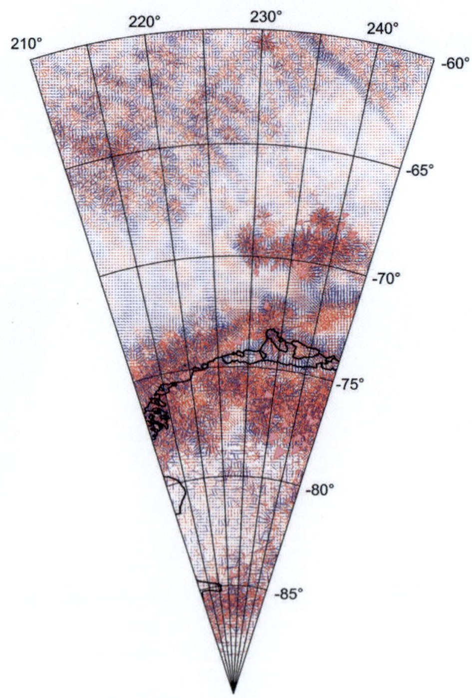

Fig. 5.8.8 *vd,* dilatation

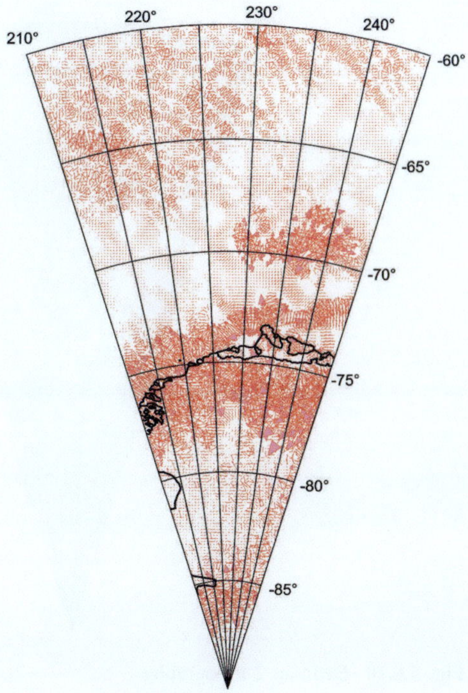

Fig. 5.8.9 *vd,* compression

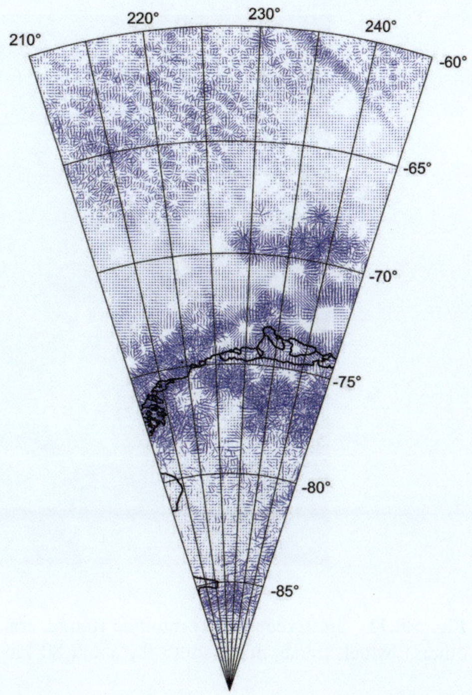

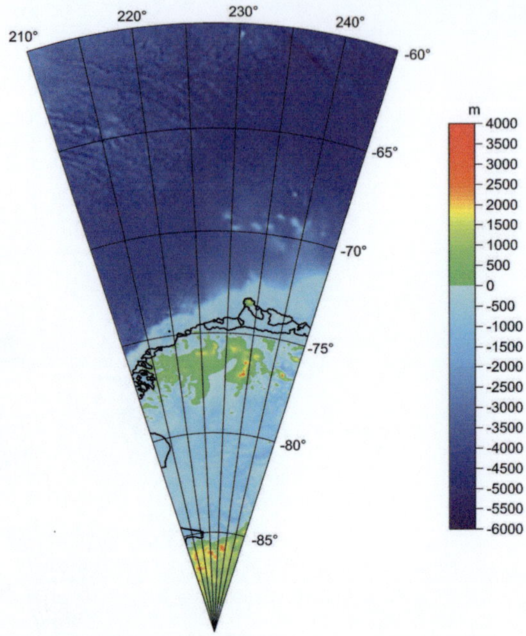

Fig. 5.8.10 Bedmap 2 topography

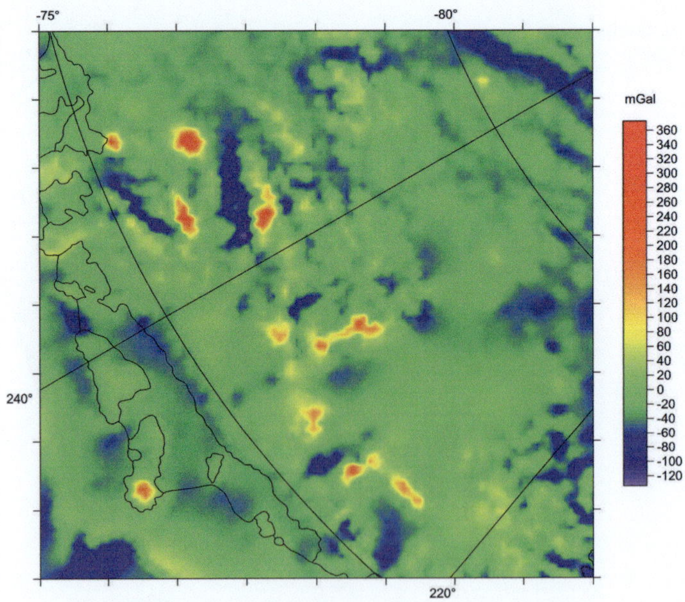

Fig. 5.8.11 *Δg*. Executive Committee Range, consisting of five major volcanoes (Mt Sidley and others), which trends north-south for about 80 km in Marie Byrd Land

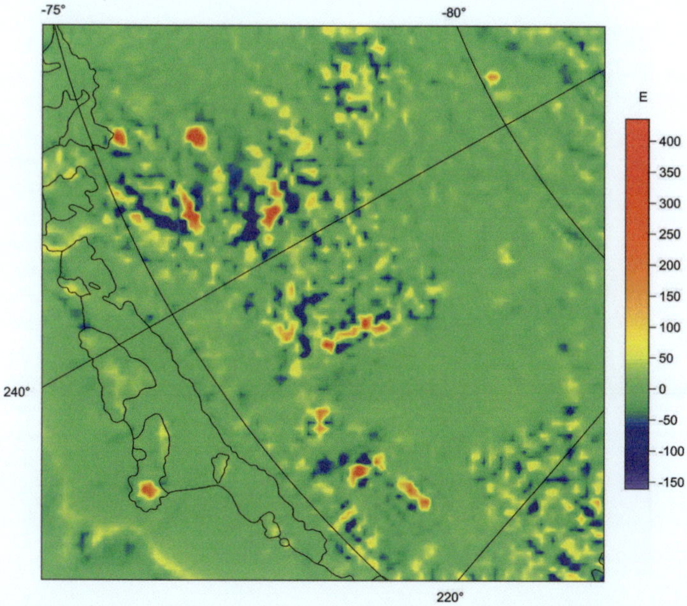

Fig. 5.8.12 T_{zz}

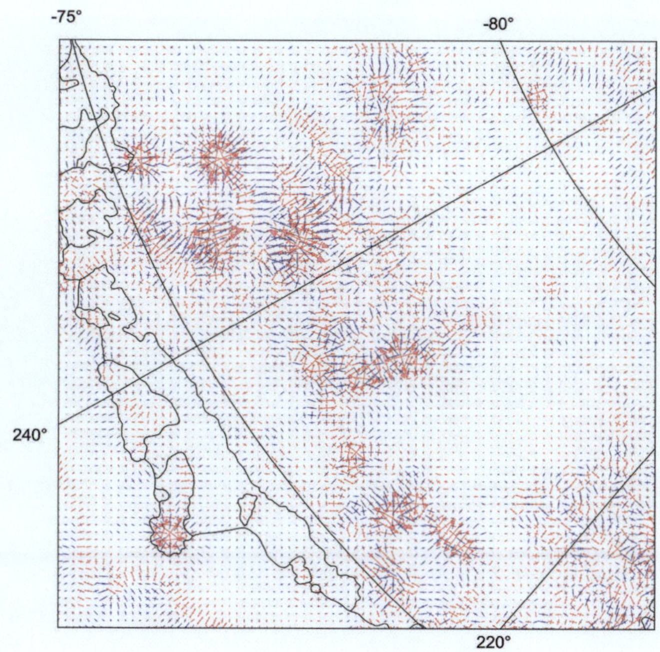

Fig. 5.8.13 *vd*

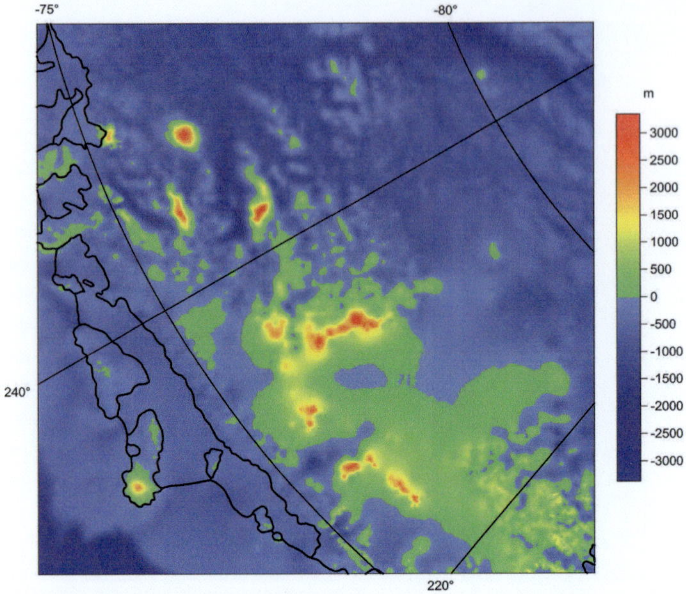

Fig. 5.8.14 Topography (Bedmap 2) in Marie Byrd Land with the Executive Committee Range

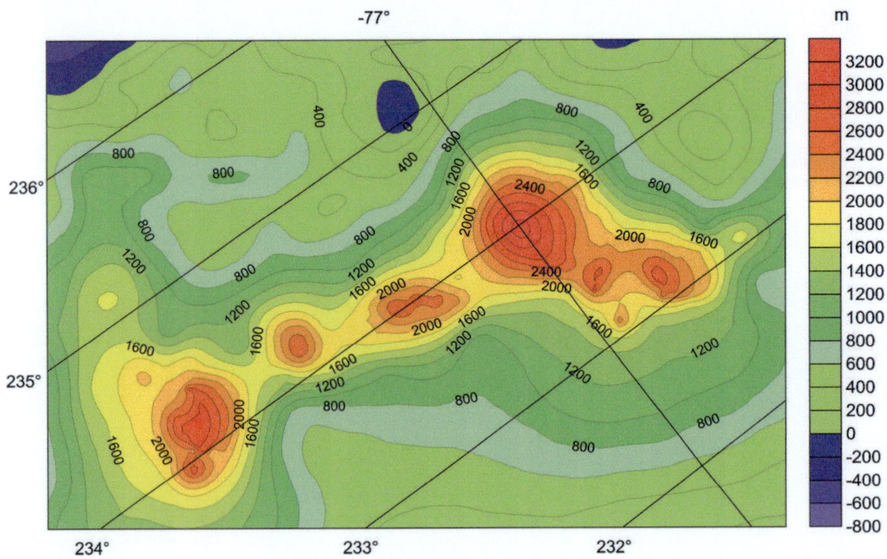

Fig. 5.8.15 Topography (Bedmap 2)—a zoom for Mt Sidley

Fig. 5.8.16 Topography (not Bedmap 2)—a zoom for Mt Sidley © Google Earth

Fig. 5.9.0 An orientation rose and introductory figure for this segment

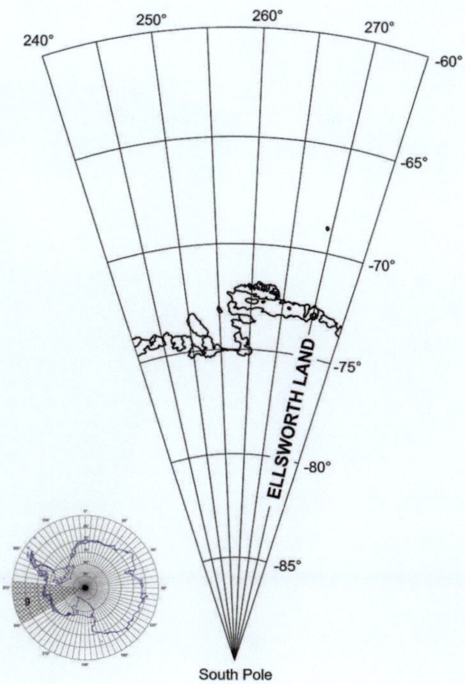

Fig. 5.9.1 Δg

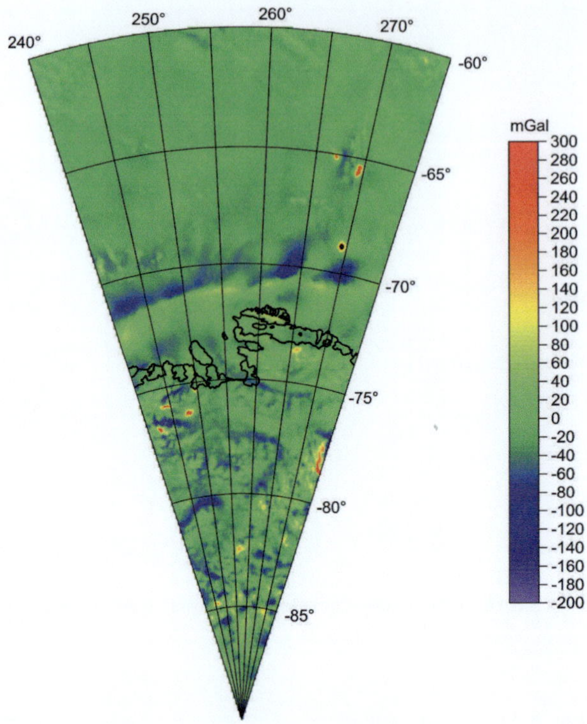

Fig. 5.9.2 T_{zz}

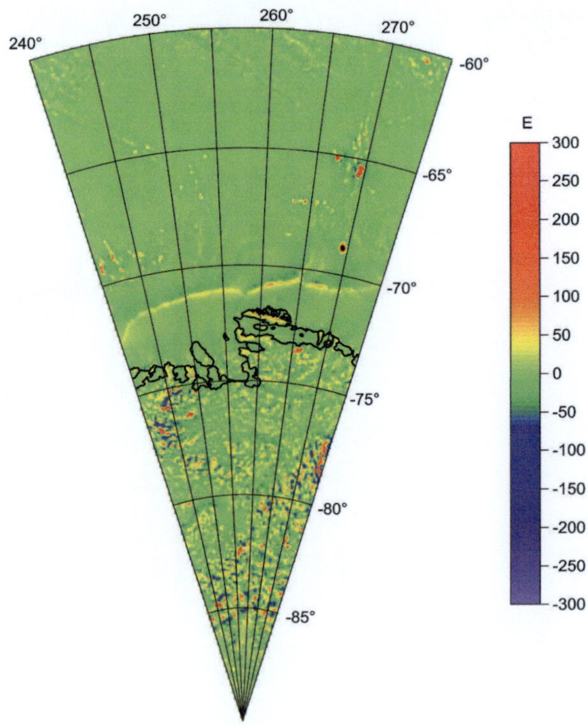

Fig. 5.9.3 I_1

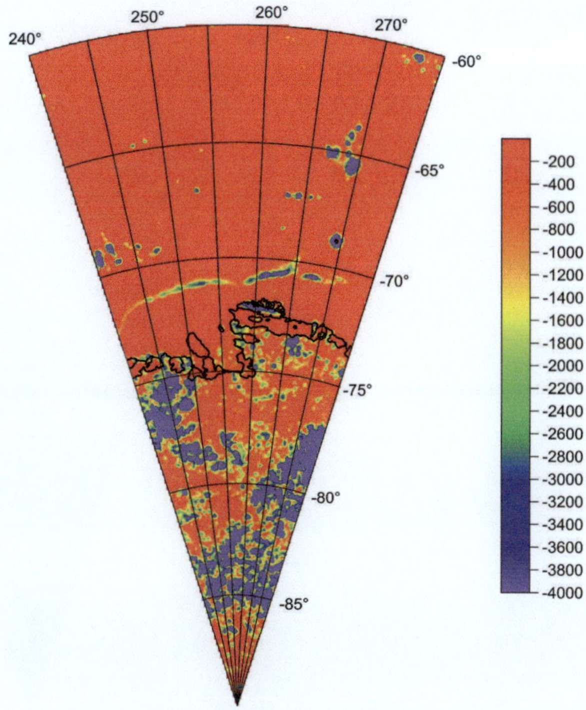

Fig. 5.9.4 I_2

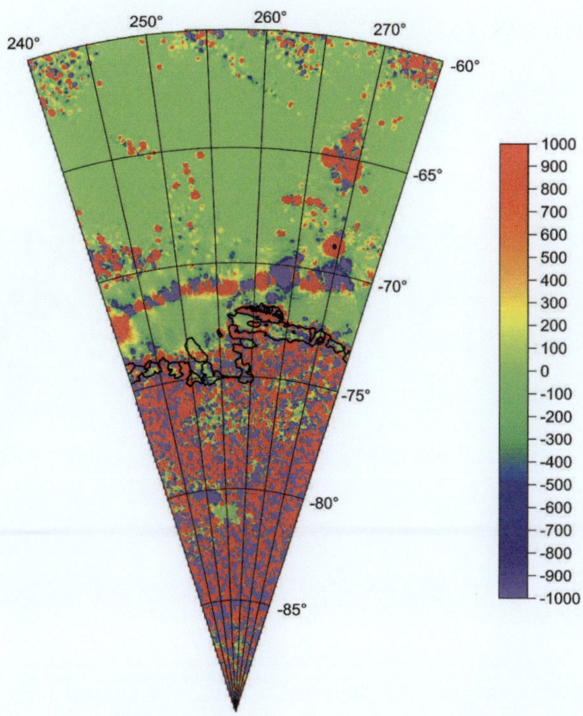

Fig. 5.9.5 I

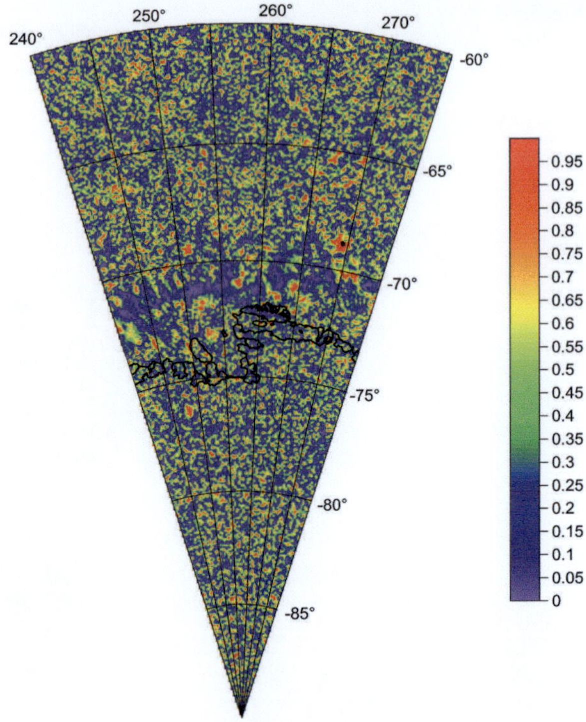

Fig. 5.9.6 θ

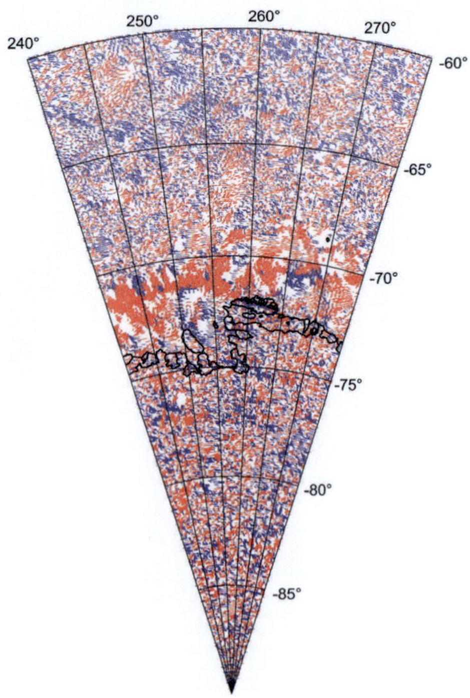

Fig. 5.9.7 *vd*

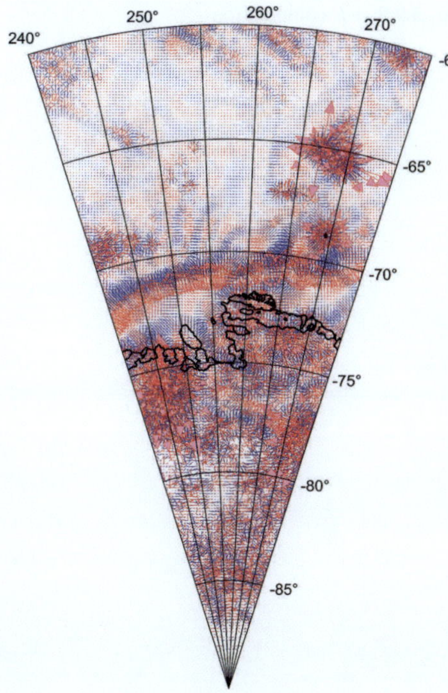

Fig. 5.9.8 *vd,* dilatation

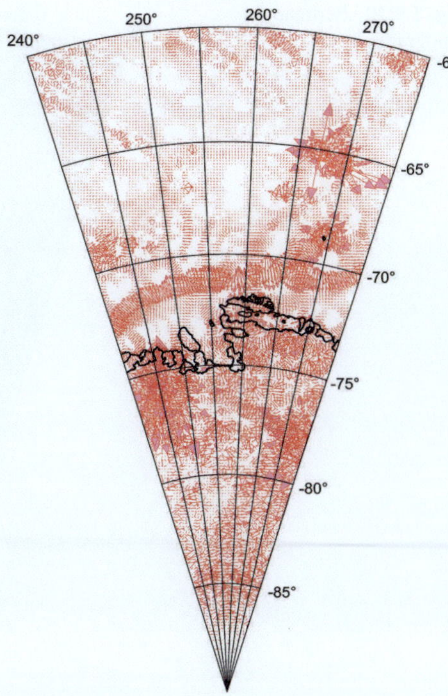

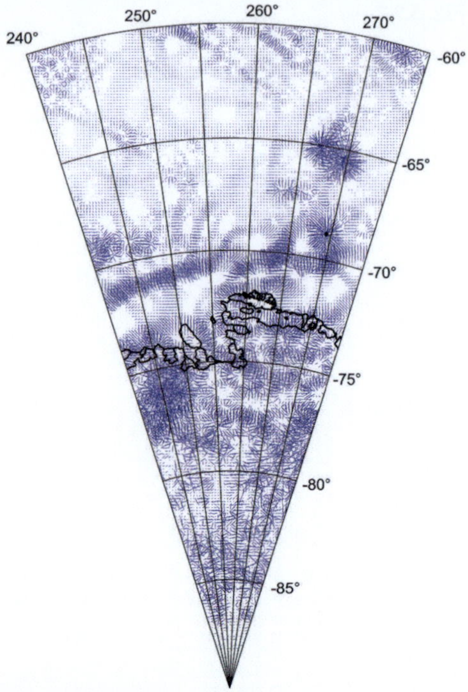

Fig. 5.9.9 *vd,* compression

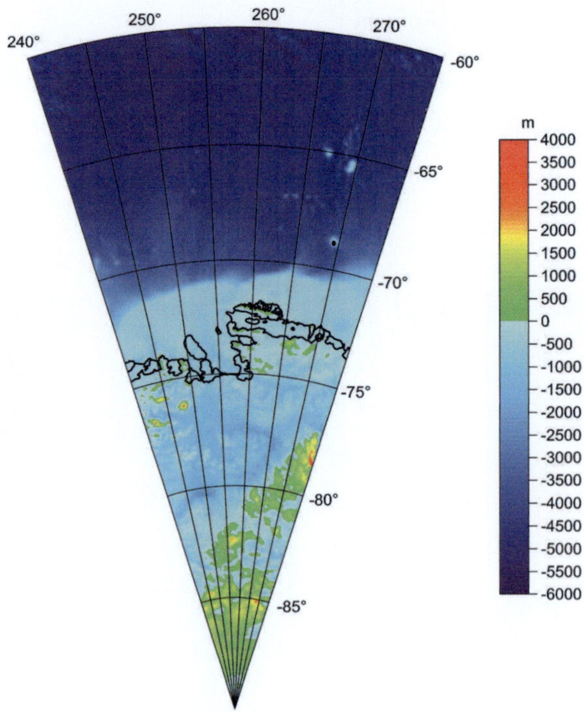

Fig. 5.9.10 Bedmap 2 topography

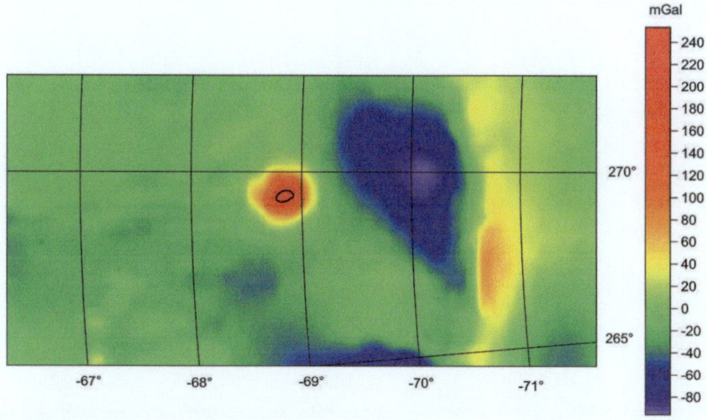

Fig. 5.9.11 Δg on the area of Peter I Island and the undersea structure north of it in the Bellingshausen Sea, near Ellsworth Land

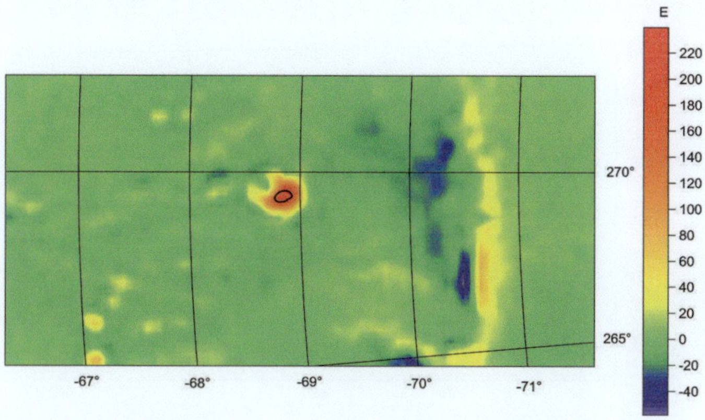

Fig. 5.9.12 T_{zz}

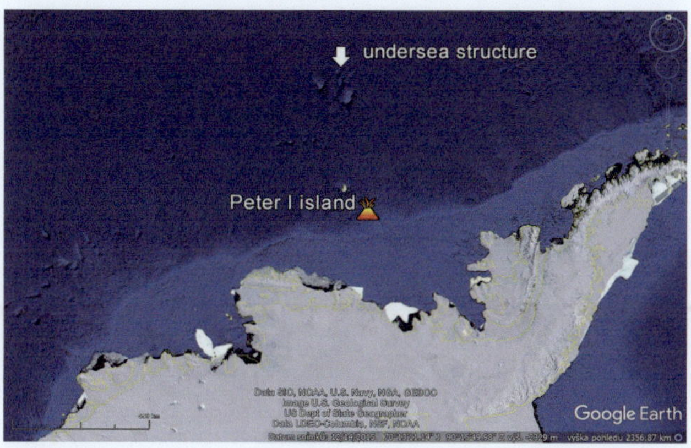

Fig. 5.9.13 © Google Earth satellite's view on the area of Peter I Island and the undersea structure north of it in the Bellingshausen Sea, near Ellsworth Land

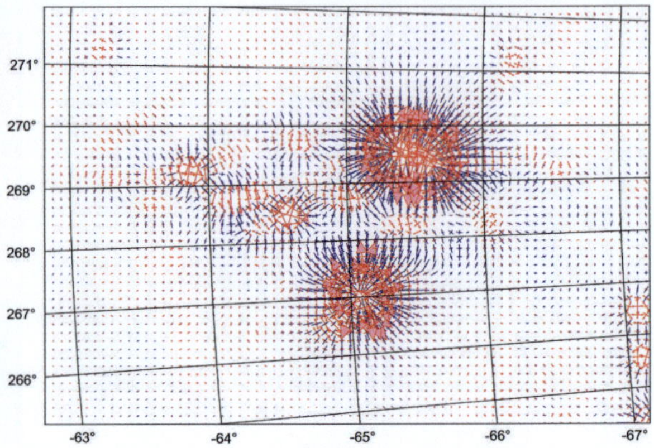

Fig. 5.9.14 The virtual deformations *vd* complete (*the upper figure*)

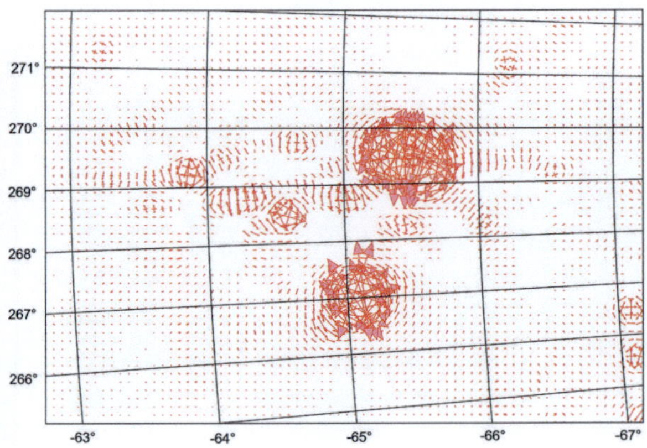

Fig. 5.9.15 The virtual deformations *vd*—dilatation

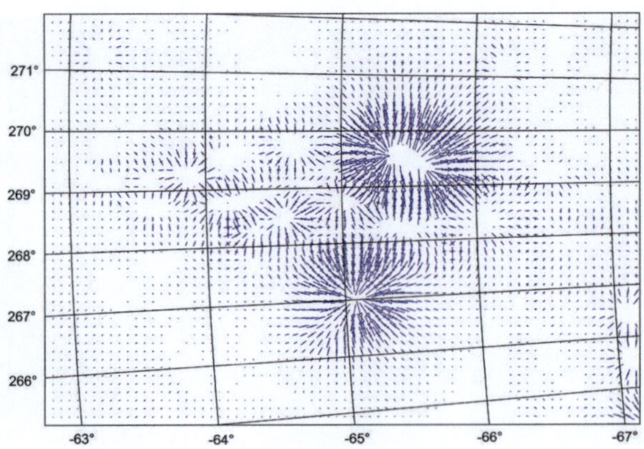

Fig. 5.9.16 The virtual deformations *vd*—compression

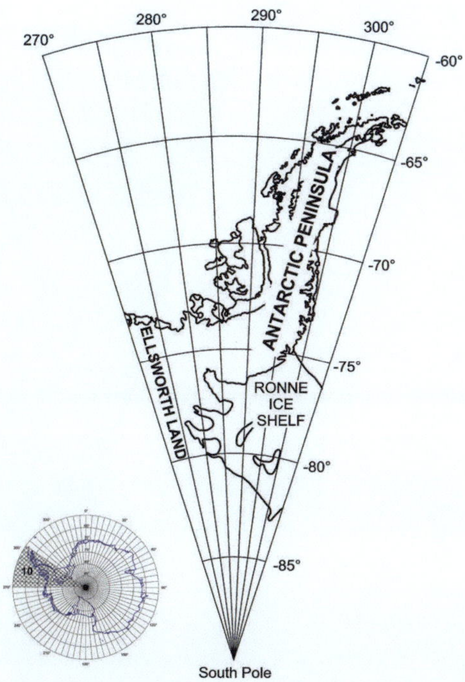

Fig. 5.10.0 An orientation rose and introductory figure for this segment showing a few topographic features of the area

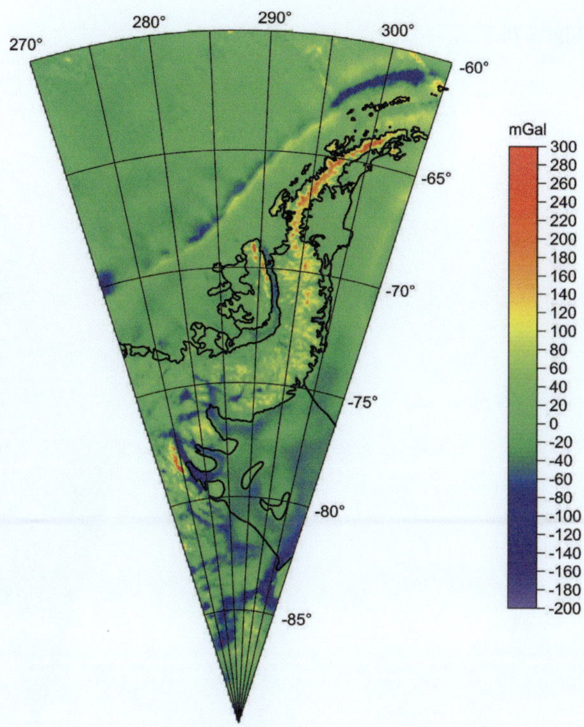

Fig. 5.10.1 Δg

Fig. 5.10.2 T_{zz}

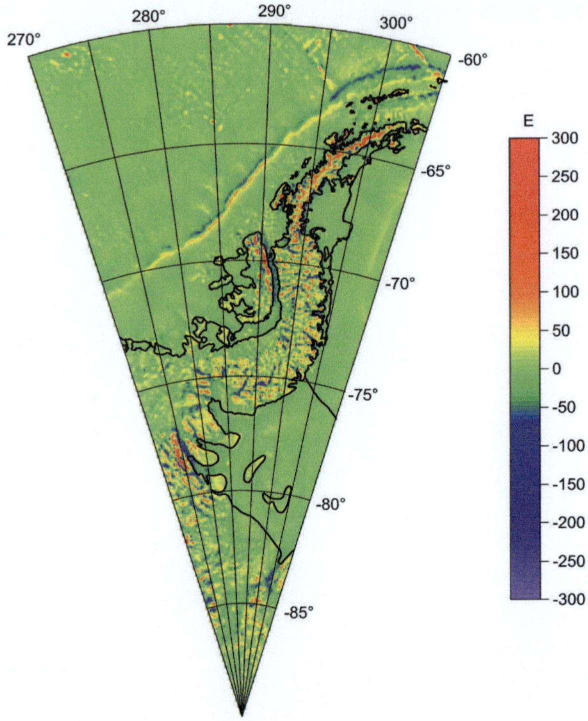

Fig. 5.10.3 I_1

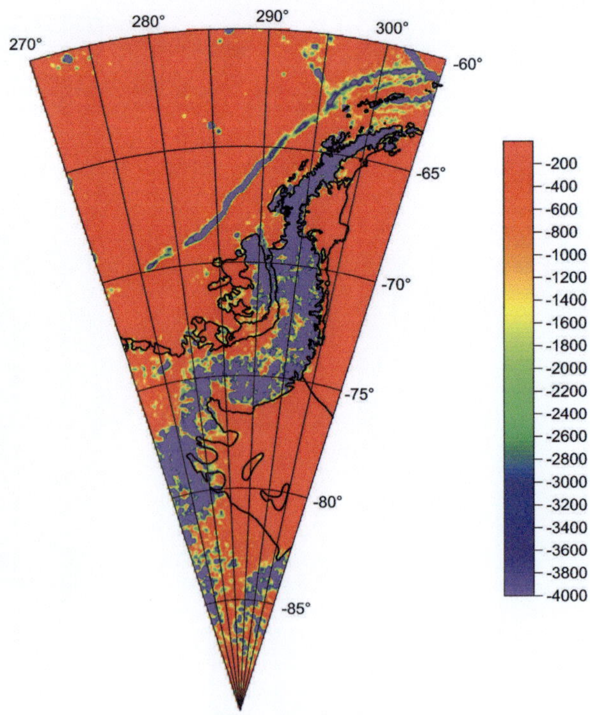

Fig. 5.10.4 I_2

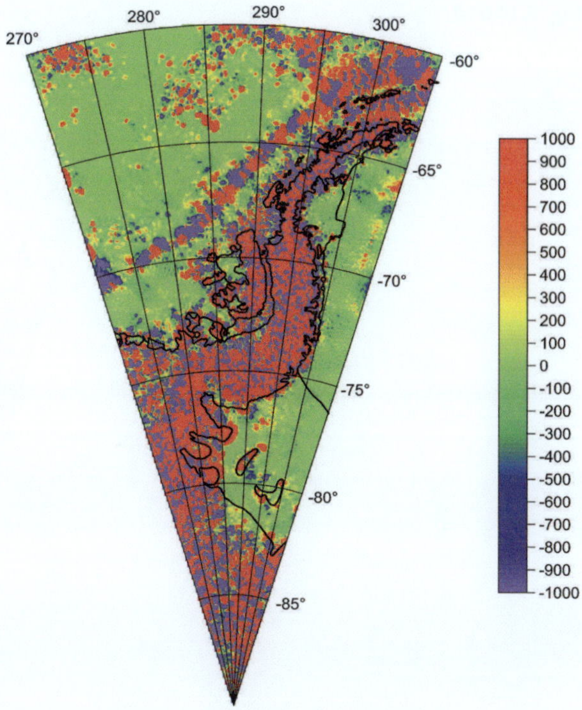

Fig. 5.10.5 I

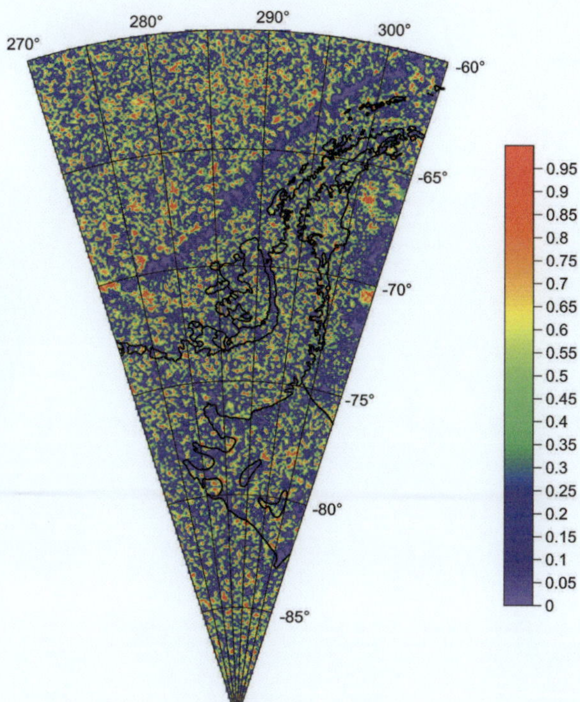

Fig. 5.10.6 θ

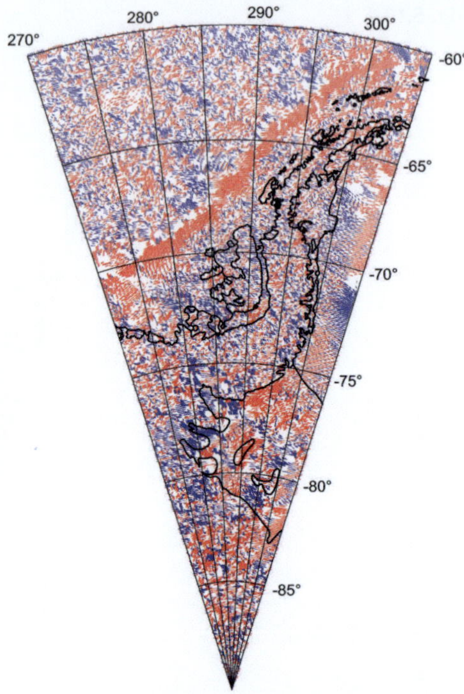

Fig. 5.10.7 *vd*

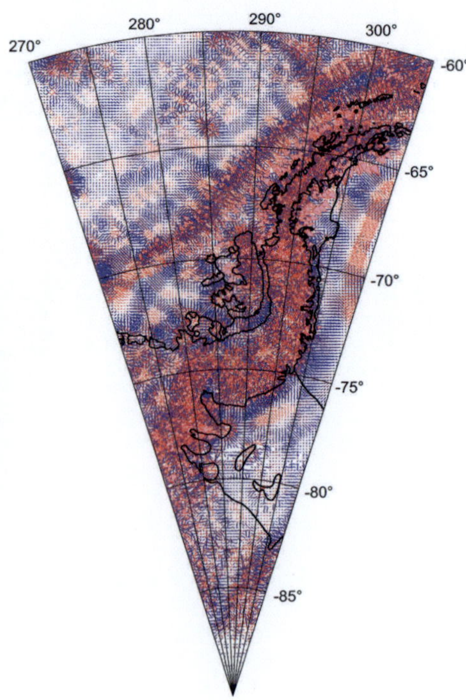

Fig. 5.10.8 *vd,* dilatation

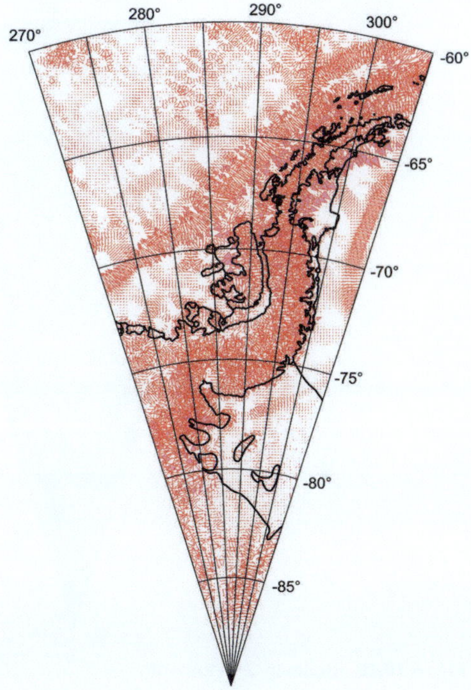

Fig. 5.10.9 *vd,* compression

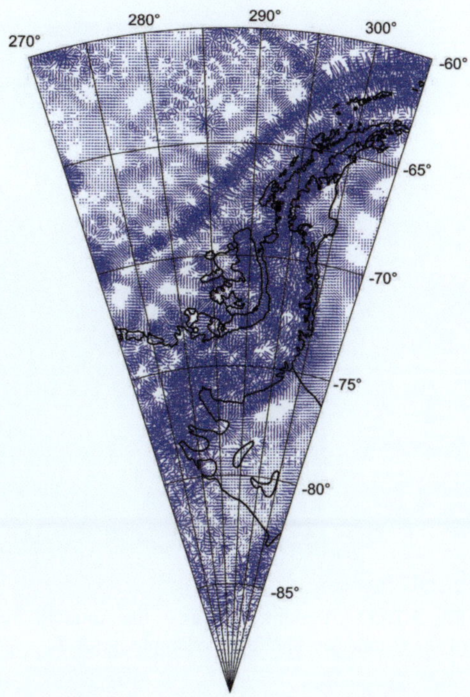

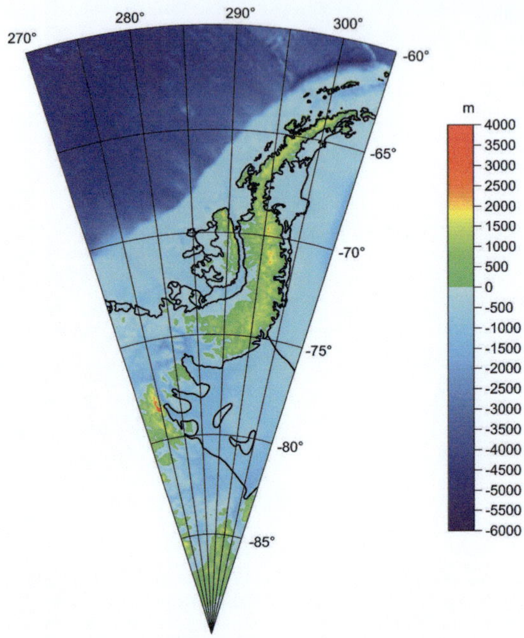

Fig. 5.10.10 Bedmap 2 topography

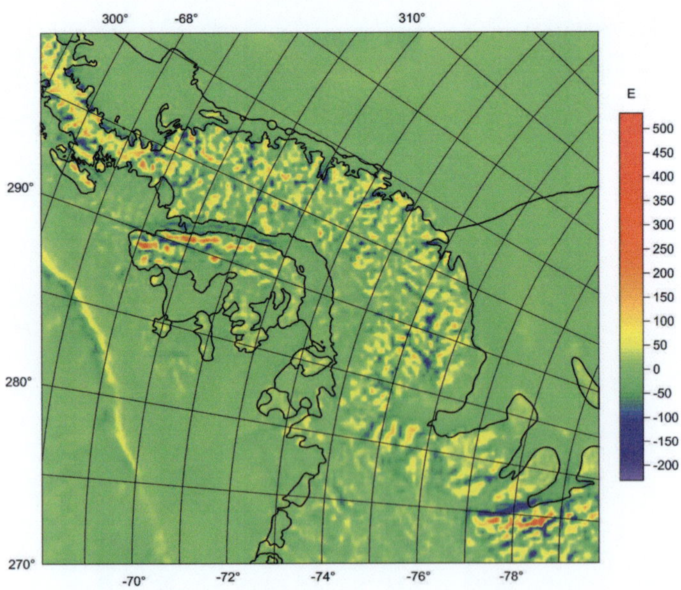

Fig. 5.10.11 A zoom of a part of the Antarctic Peninsula, Weddell Sea, Palmer Land, and a part of the Ronne Ice Shelf; the example is for T_{zz}

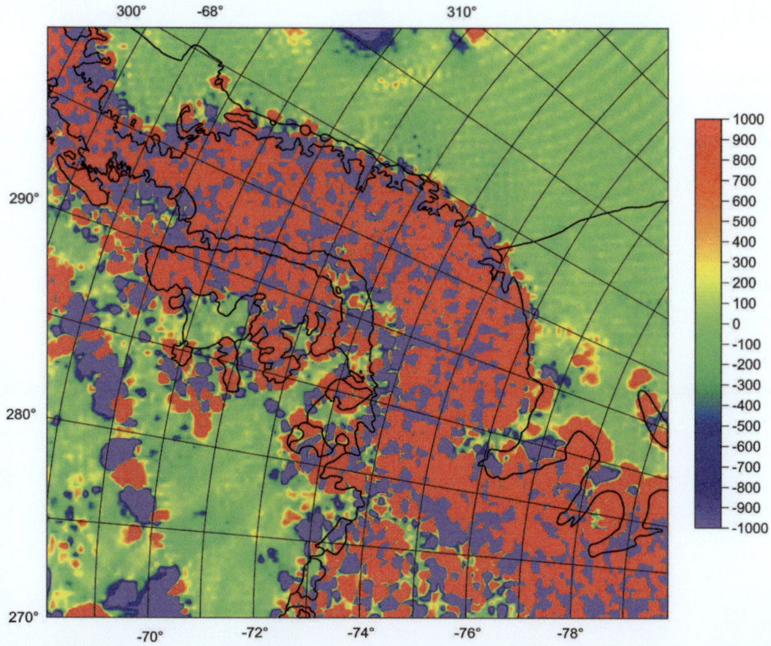

Fig. 5.10.12 A zoom of a part of the Antarctic Peninsula, Weddell Sea, Palmer Land, and a part of the Ronne Ice Shelf; this example is for I_2

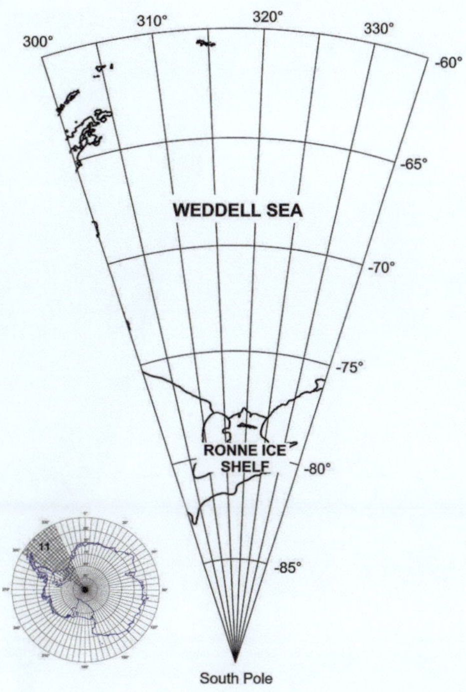

Fig. 5.11.0 An orientation rose and introductory figure for this segment

Fig. 5.11.1 Δg

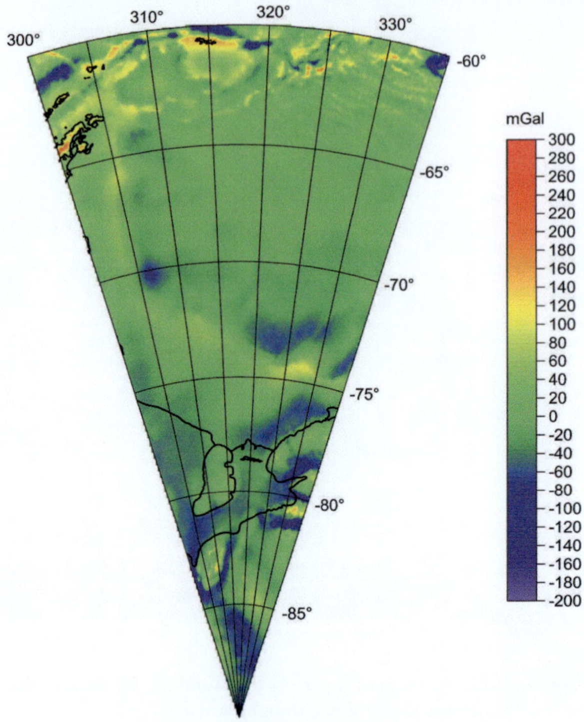

Fig. 5.11.2 T_{zz}

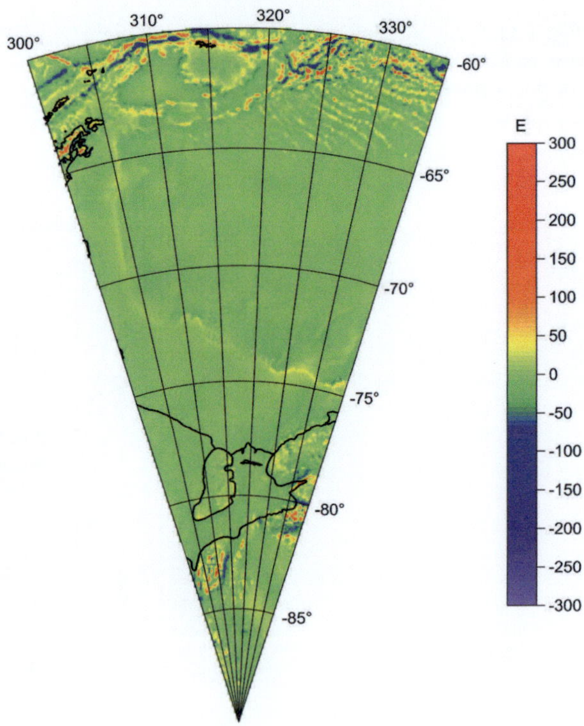

Fig. 5.11.3 I_1

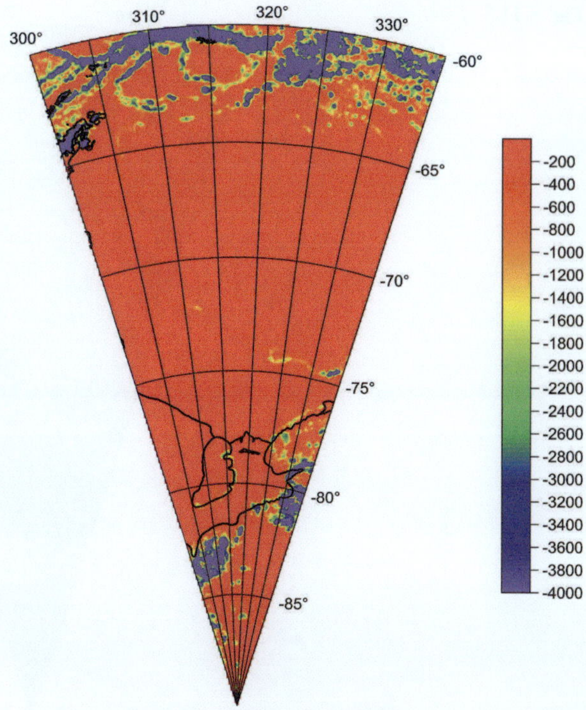

Fig. 5.11.4 I_2

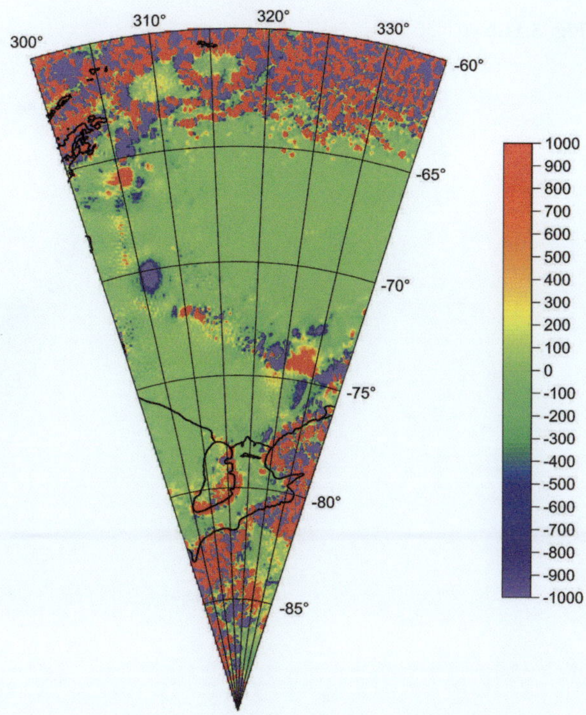

Fig. 5.11.5 I

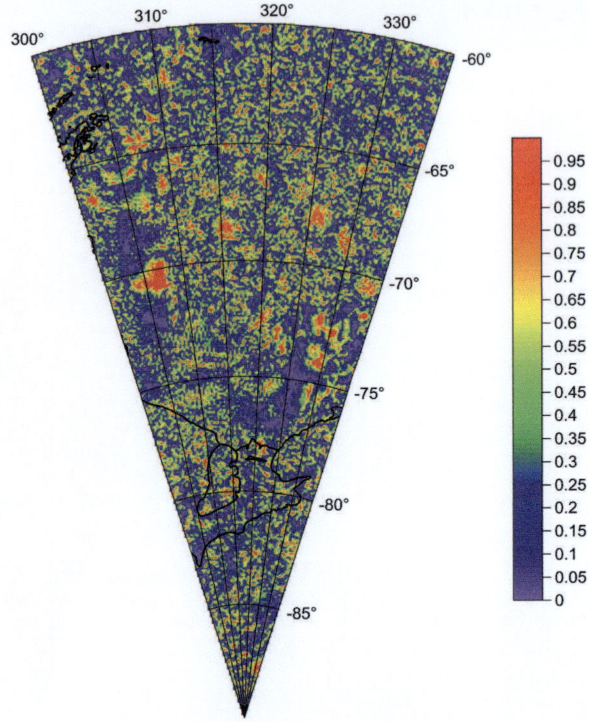

Fig. 5.11.6 θ

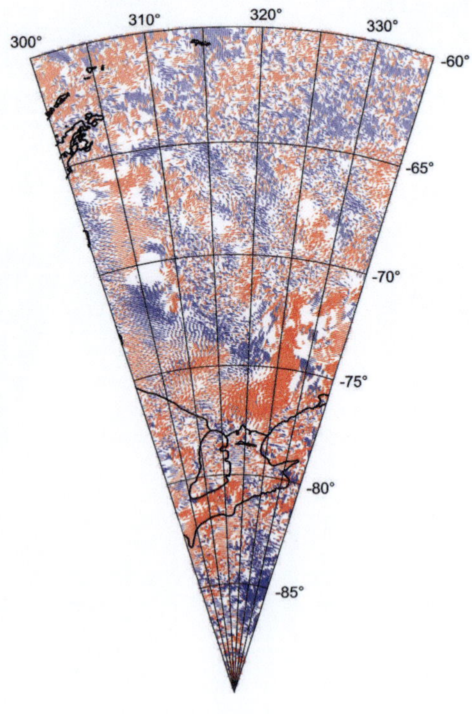

Fig. 5.11.7 *vd*

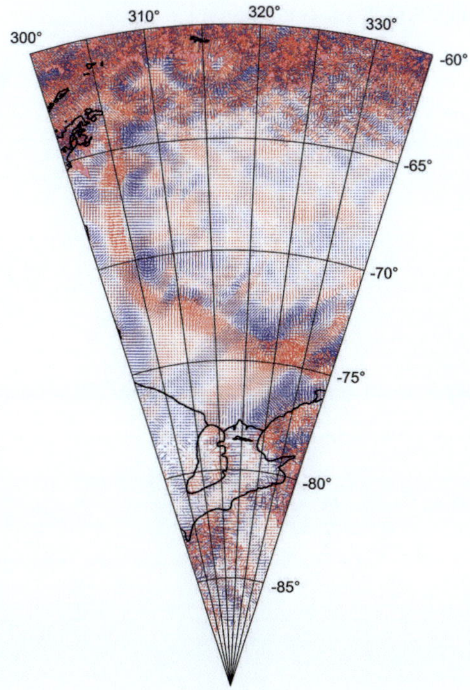

Fig. 5.11.8 *vd,* dilatation

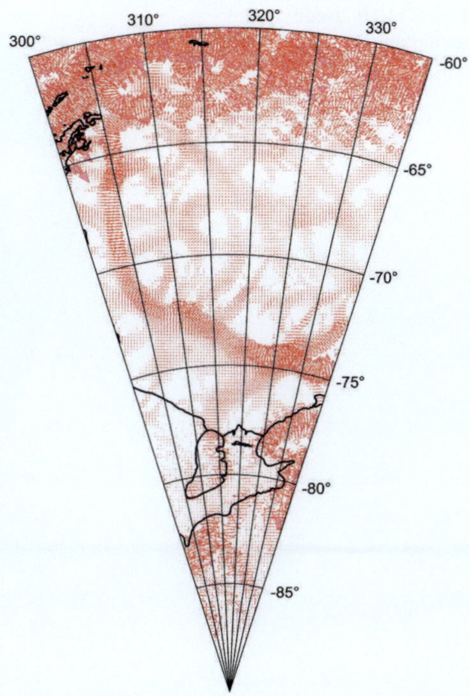

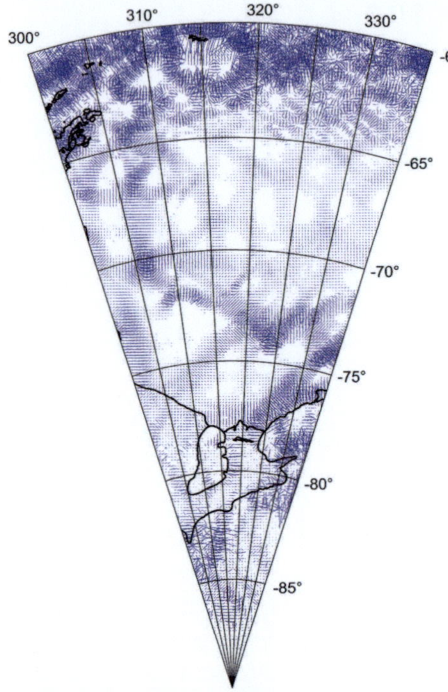

Fig. 5.11.9 *vd,* compression

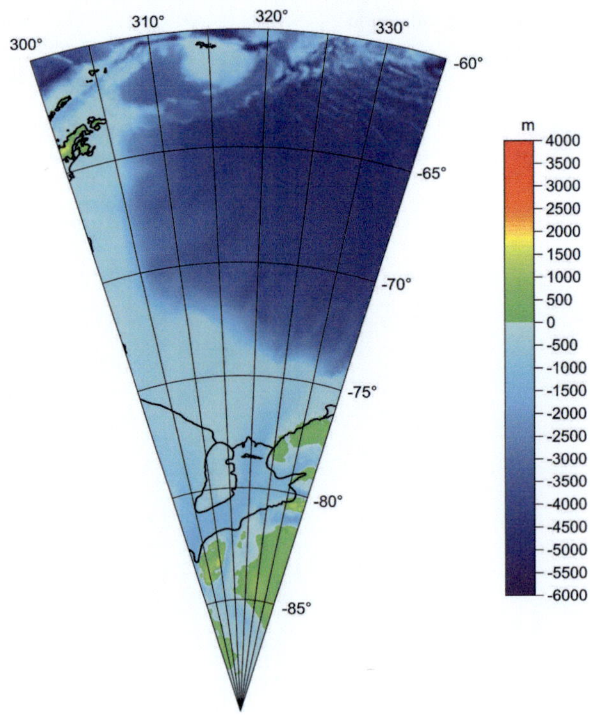

Fig. 5.11.10 Bedmap 2 topography

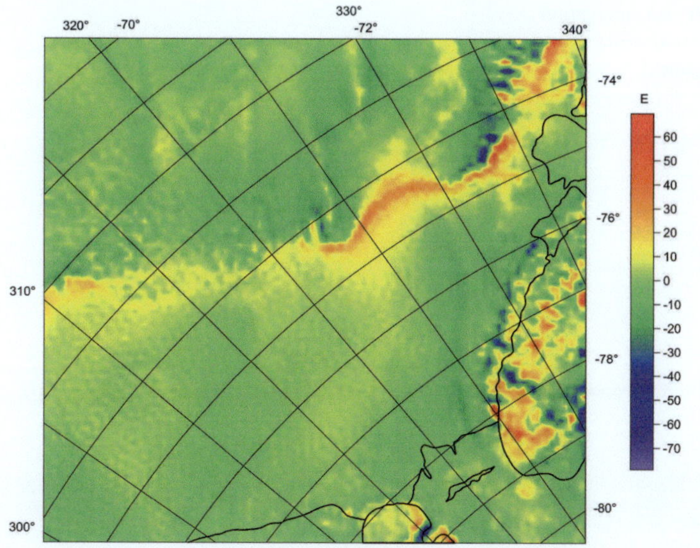

Fig. 5.11.11 A close-up to the Weddell Sea and the Ronne Ice Shelf, T_{zz} (*the upper figure*)

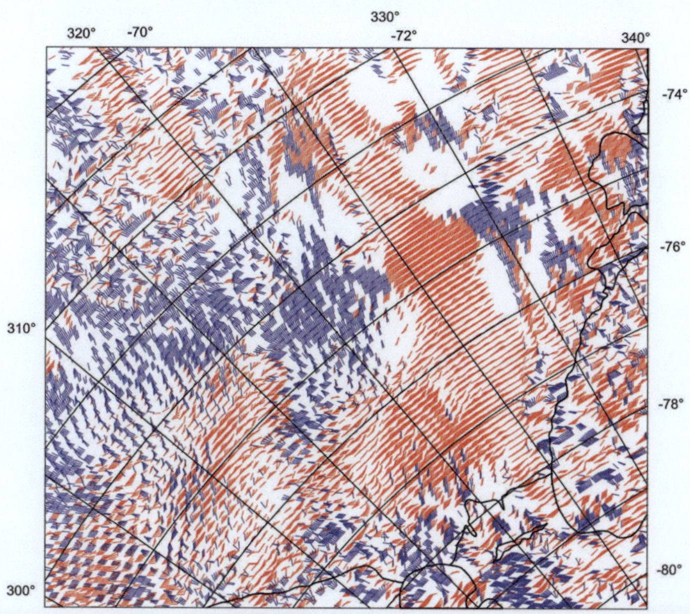

Fig. 5.11.12 A close-up to the Weddell Sea and the Ronne Ice Shelf, the strike angle θ (vectors "combed" into one prevailing direction; similar situation is described in Klokočník and Kostelecký 2014)

Fig. 5.12.0 An orientation rose and introductory figure for this segment

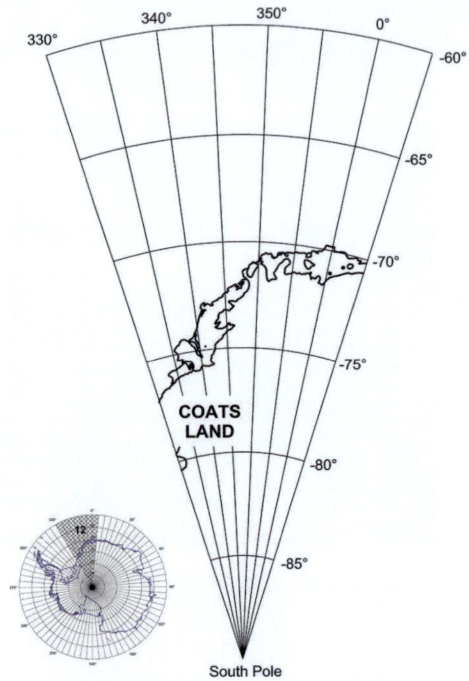

Fig. 5.12.1 Δg

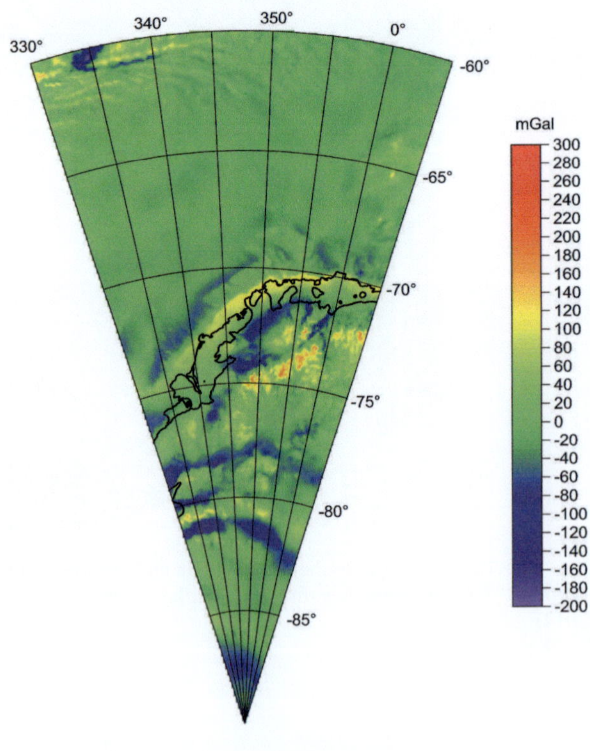

Fig. 5.12.2 T_{zz}

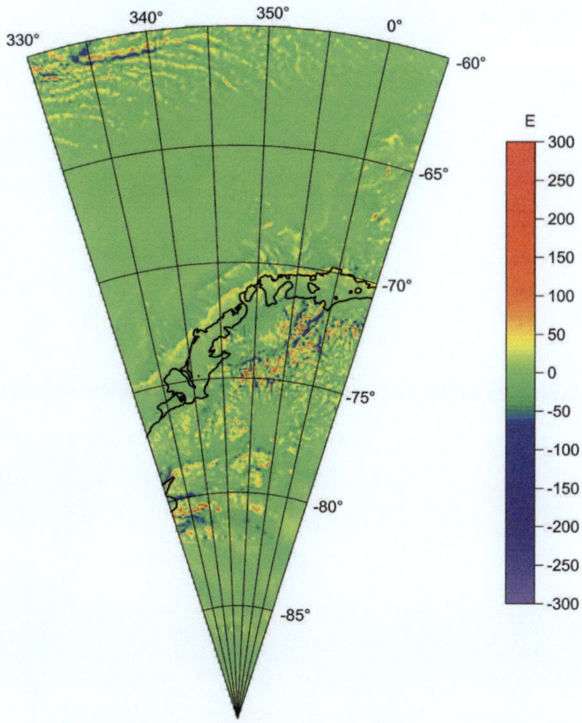

Fig. 5.12.3 I_1

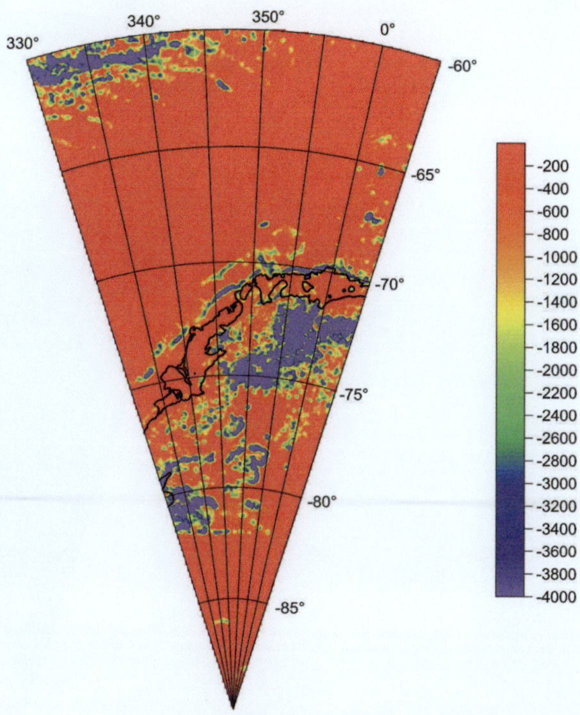

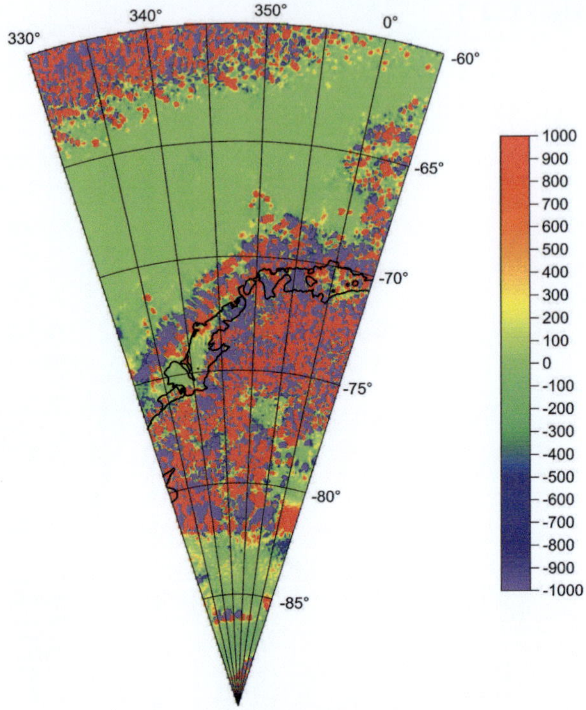

Fig. 5.12.4 I_2

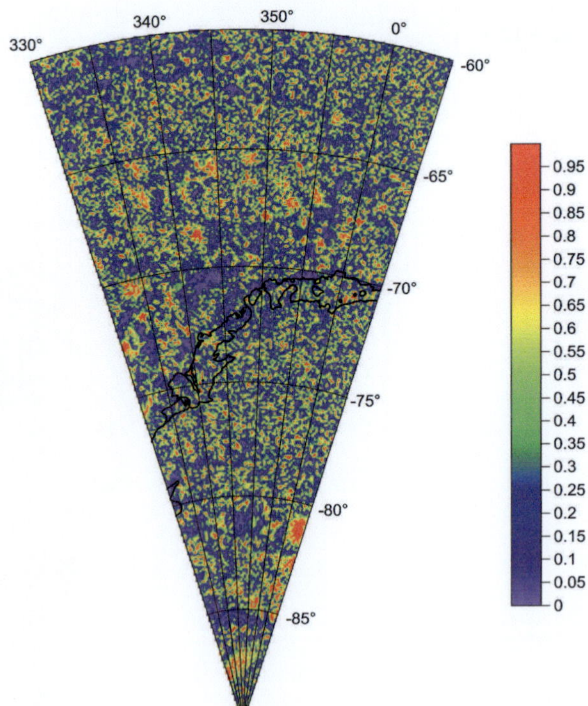

Fig. 5.12.5 I

Fig. 5.12.6 θ

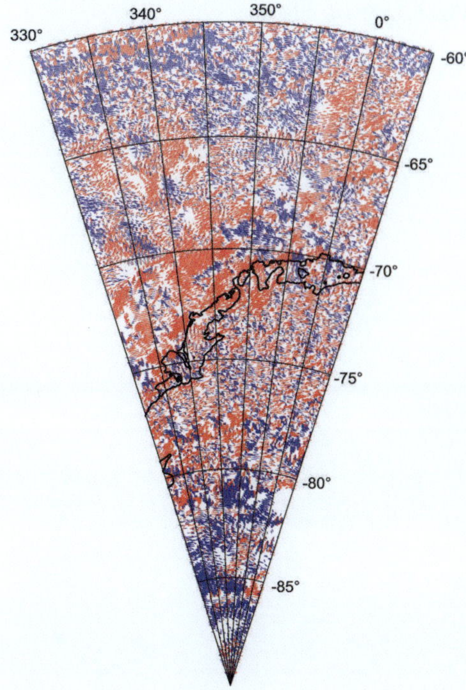

Fig. 5.12.7 *vd*

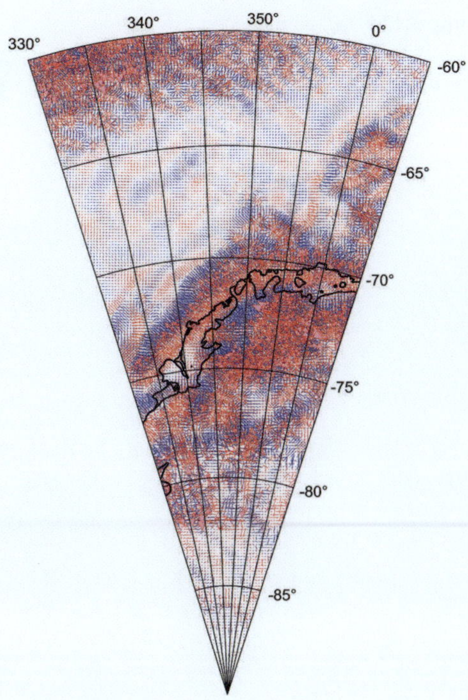

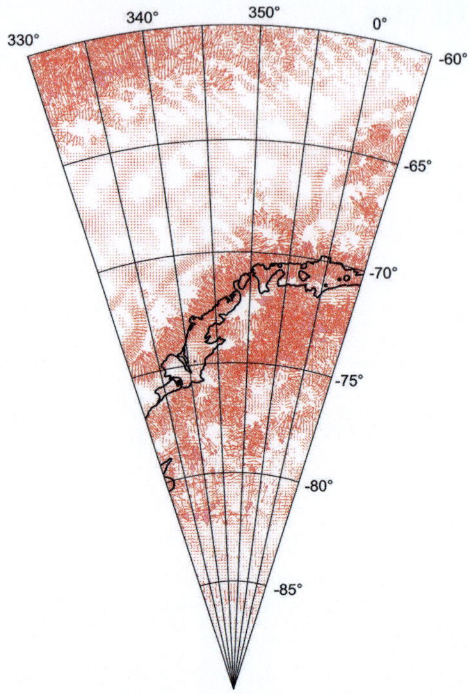

Fig. 5.12.8 *vd,* dilatation

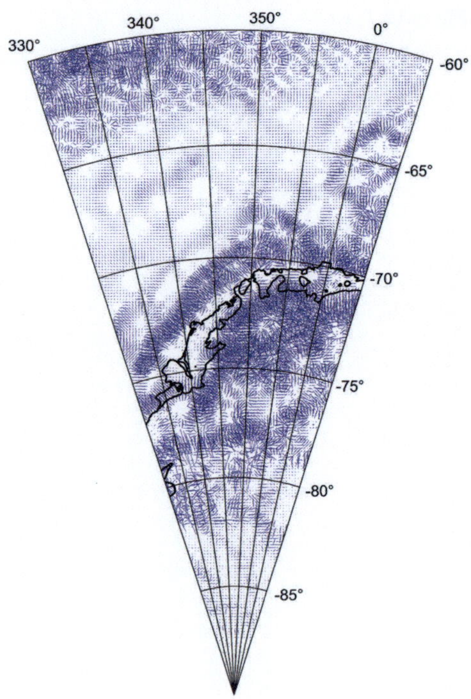

Fig. 5.12.9 *vd,* compression

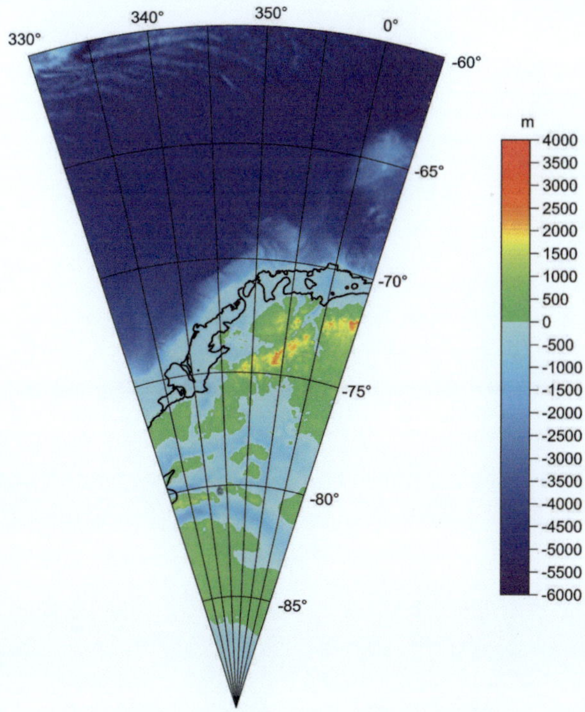

Fig. 5.12.10 Bedmap 2 topography

References

Klokočník J. and Kostelecký J. 2014. Gravity signal at Ghawar, Saudi Arabia, from the global gravitational field model EGM 2008 and similarities around, *Arab. J. Geosciences*, on-line, doi: 10.1007/s12517-014-1491-y; ISSN 1866-7511, Springer-Verlag.

Klokočník J., Kostelecký J., Bezděk A. and Pešek I. 2016. On feasibility to detect volcanoes hidden under the ice of Antarctica via their "gravitational signal", *Annals of Geophysics* 59(5), S0539. http://dx.doi.org/10.4401/ag-7102.

MIX
Papier aus verantwortungsvollen Quellen
Paper from responsible sources
FSC® C105338

If you have any concerns about our products,
you can contact us on
ProductSafety@springernature.com

In case Publisher is established outside the EU,
the EU authorized representative is:
Springer Nature Customer Service Center GmbH
Europaplatz 3, 69115 Heidelberg, Germany

Printed by Libri Plureos GmbH
in Hamburg, Germany